Analog/Digital Implementation of Fractional Order Chaotic Circuits and Applications

Esteban Tlelo-Cuautle • Ana Dalia Pano-Azucena
Omar Guillén-Fernández • Alejandro Silva-Juárez

Analog/Digital Implementation of Fractional Order Chaotic Circuits and Applications

Springer

Esteban Tlelo-Cuautle
INAOE
Tonantzintla
Puebla, Mexico

Ana Dalia Pano-Azucena
INAOE
Tonantzintla
Puebla, Mexico

Omar Guillén-Fernández
INAOE
Tonantzintla
Puebla, Mexico

Alejandro Silva-Juárez
INAOE
Tonantzintla
Puebla, Mexico

ISBN 978-3-030-31252-7   ISBN 978-3-030-31250-3   (eBook)
https://doi.org/10.1007/978-3-030-31250-3

This Springer imprint is published by the registered company Springer Nature Switzerland AG.
The registered company address is: Gewerbestrasse 11, 6330 Cham, Switzerland

Esteban Tlelo-Cuautle, Ana Dalia Pano-Azucena, Omar Guillén-Fernández, and Alejandro Silva-Juárez want to thank to their families for the support provided during the time spent writing this book.

Preface

Nowadays, noninteger or fractional-order chaotic systems are very interesting topics to engineers, physicists, and mathematicians because most real physical systems are inherently nonlinear in nature. It is worth mentioning that several books have recently been published detailing implementations of integer-order chaotic and hyperchaotic circuits and systems using analog integrated circuits technology, discrete devices, field-programmable analog arrays (FPAAs), and digital hardware like micro-controllers, field-programmable gate arrays (FPGAs), and multi-core systems. However, this is not the case for fractional-order chaotic systems, which mathematical models consist of fractional derivatives in nonlinear equations that are difficult to be solved by analytical methods.

Recent works have demonstrated that fractional-order differential equations are, at least, as stable as their integer-order counterpart. So that one can study a fractional-order mathematical model of a chaotic system to evaluate their equilibrium points, stability, and perform numerical simulation applying appropriate methods in the frequency and time domains, such as Caputo's definition, Grünwald–Letnikov method, and Adams-Bashforth-Moulton method. In addition, and under certain conditions, the dynamics analyses required by fractional-order chaotic systems can be performed in the same way as those done for integer-order ones, e.g., evaluation of the Lyapunov exponents, entropy, and fractal dimension.

Integer-order chaotic systems are relatively simple to implement with analog and digital hardware. However, analog implementations suffer the sensitivity problem of the analog component values and digital implementations suffer the problem of degradation due to the reduced number of bits to perform computer arithmetic operations. These problems are more difficult to solve in implementing fractional-order chaotic systems, which require more complex analog circuitry to solve fractional derivatives, thus increasing variations in the circuit elements; and require more complex digital hardware to implement memory blocks and control blocks that are required by time simulation methods. In this manner, it can be appreciated that the main objective of recent and related scientific works are oriented to introduce alternatives for the analog/digital implementation of fractional-order chaotic oscillators, trying to provide details on the synthesis and physical realization using either analog or digital electronics. That way, this book provides guidelines for analog and digital implementations of fractional-order chaotic systems, which are performed from applying numerical methods in the frequency and time domains, to approximate the solutions and to synthesize the mathematical descriptions using amplifiers, FPAAs and FPGAs. The electronic implementations are measured in laboratory conditions to observe experimental fractional-order chaotic attractors, which are used to implement random number generators and secure communication systems for image encryption.

The contents of this book are organized in seven chapters devoted to describe the basic theory and review of mathematical models of several fractional-order chaotic systems that are simulated with different numerical methods; a review on the implementation of integer-/fractional-order chaotic oscillators using analog/digital hardware; characterization of the Lyapunov exponents, Kaplan–Yorke dimension and entropy, and their optimization applying metaheuristics; guidelines to approximate fractional derivatives in the frequency domain and its implementation using amplifiers and FPAAs; guidelines to perform VHDL descriptions of different fractional-order chaotic oscillators, which can be implemented on FPGAs, and guidelines to reduce hardware resources; and details to implement random number generators and the synchronization of fractional-order chaotic oscillators applying different techniques to develop applications in chaos-based secure communications and image encryption. The book ends with a discussion of recent and future trends on this hot topic from fractional calculus and highlights alternatives on the electronic implementation of fractional-order chaotic circuits and systems to enhance modern engineering applications in the era of hardware security for Internet of Things.

INAOE, Tonantzintla, Mexico   Esteban Tlelo-Cuautle
INAOE, Tonantzintla, Mexico   Ana Dalia Pano-Azucena
INAOE, Tonantzintla, Mexico   Omar Guillén-Fernández
INAOE, Tonantzintla, Mexico   Alejandro Silva-Juárez

Acknowledgment

Ana Dalia Pano-Azucena, Omar Guillén-Fernández, and Alejandro Silva-Juárez want to thank to CONACyT for their scholarships to pursue a Doctoral degree in Electronics Science, and to INAOE for the academic teaching, scientific research training, and all infrastructure and support provided during their postgraduate studies.

Contents

Acronyms

CAB	Configurable Analog Block
CAM	Configurable analog module
CLB	Configurable logic block
CLK	Clock
CMOS	Complementary metal-oxide-semiconductor
D_{KY}	Kaplan–Yorke dimension
DAC	Digital-to-analog converter
DE	Differential evolution
DSP	Digital signal processor
EDA	Electronic design automation
EP	Equilibrium point
FE	Forward Euler
FPAA	Field-programmable analog array
FPGA	Field-programmable gate array
FSM	Finite state machine
h	Step-size
IEEE	Institute of Electrical and Electronics Engineers
IoT	Internet of Things
IP	Intellectual property
LE+	Positive Lyapunov exponent
LUT	Lookup table
MatLab	Matrix Laboratory
ODE	Ordinary differential equation
OpAmp	Operational amplifier
OPCL	Open-Plus-Closed-Loop
OTA	Operational transconductance amplifier
PID	Proportional-integral-derivative
PSO	Particle swarm optimization
PVT	Process-Voltage-Temperature
PWL	Piecewise-linear
RAM	Random access memory

RK	Runge–Kutta
ROM	Read only memory
RST	Reset
SCM	Single constant multiplier
SNLF	Saturated nonlinear function
TISEAN	Time Series Analyzer
VHDL	Very-high-speed-integrated-circuit hardware description language

Chapter 1
Integer and Fractional-Order Chaotic Circuits and Systems

1.1 Chaotic Circuits and Systems

In general, a nonlinear system is a system which is not linear, that is, a system which does not satisfy the superposition principle. For instance, the authors in [1] state that a nonlinear system is any problem, where the variables to be solved cannot be written as a linear combination of independent components. If the equation contains a nonlinear function (power or cross product), the system is nonlinear as well. Nonlinear systems are very interesting to engineers, physicists, and mathematicians because most real physical systems are inherently nonlinear in nature. Nonlinear equations are difficult to be solved by analytical methods and give rise to interesting phenomena such as bifurcation and chaos. Even simple nonlinear or piecewise-linear (PWL) dynamical systems can exhibit completely an unpredictable behavior, the so-called deterministic chaos, for which there are few definitions of the chaotic dynamics, e.g.: (1) a system with at least one positive Lyapunov exponent is chaotic; (2) a system with positive entropy is chaotic; (3) a system equivalent to hyperbolic or Anosov system is chaotic, etc. The common part of all definitions is the existence of local instability and divergence of initially close trajectories. At the same time, all definitions are not exactly equivalent.

The word chaos is generally associated with disorder, and the dynamical systems presenting this behavior have super-sensibility to the initial conditions, which makes them stochastic. Besides, chaotic systems are well defined by nonlinear equations and their behavior seems to be irregular or random, but its nature is fully deterministic, i.e., there exists a differential equation that defines its dynamics and from it one can know with good precision the data sequence. According to [2], chaotic systems have a new classification that can be performed according to their nonlinear dynamics, so that one can distinguish two kinds of attractors: self-excited and hidden attractors. In the former case, the attractor has a basin of attraction that is excited from unstable equilibrium point, as the classical nonlinear systems such as Lorenz's, Rössler's, Chen's, Lü's, or Sprott's systems.

© Springer Nature Switzerland AG 2020
E. Tlelo-Cuautle et al., *Analog/Digital Implementation of Fractional Order Chaotic Circuits and Applications*, https://doi.org/10.1007/978-3-030-31250-3_1

The other case is hidden attractors, which recently are receiving great attention from both the theoretical and practical point of views. Self-excited attractors can be localized straightforwardly yet applying manual calculations. In contrast, one must develop specific computational procedures to identify a hidden attractor because the evaluation of their equilibrium points is difficult and they do not help in their localization. This chapter deals with chaotic oscillators generating self-excited attractors, which models are described by differential equations that generate chaotic behavior [3].

Chaotic behavior is defined as a long-term aperiodic behavior of a deterministic system that exhibits dependence and becomes extremely sensitive to initial conditions. Basically, a chaotic system possesses the following characteristics [4]: High sensitivity to initial conditions; deterministic characteristic because their parameters are known; as time increases ($t \rightarrow \infty$) its behavior is difficult to distinguish, and the trajectories of the state variables do not tend to a fixed point, periodic, or quasi-periodic orbits; the strange attractors have a fractal dimension; and it has at least one positive Lyapunov exponent. The majority of authors agree that the most important characteristic is the high sensitive dependence to initial conditions, which means that observing one state from two initial conditions with infinitesimal difference, the resulting state trajectories will diverge exponentially as time increases but the range of values is bounded. Figure 1.1 shows this characteristic when simulating Lorenz chaotic oscillator with two different initial conditions for the state variable x, which mathematical model is given in (1.1). As it can be appreciated, as time increases the trajectories diverge exponentially while the difference between the two initial conditions is small as 1×10^{-6}, and this is why one calls an unpredictable chaotic behavior complex [5].

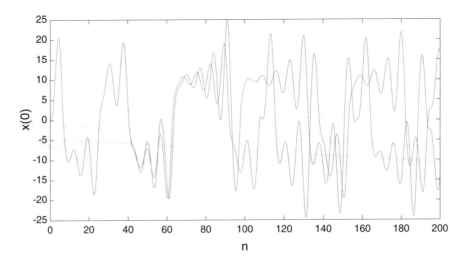

Fig. 1.1 Simulation of state variable x of Lorenz chaotic oscillator using two initial conditions: (**a**) $x(0) = 0.1$ (in blue) and (**b**) $x(0) = 0.100001$ (in red)

$$\dot{x} = \sigma(x - y)$$
$$\dot{y} = x(\rho - z) - y \qquad (1.1)$$
$$\dot{z} = xy - \beta z$$

Other well-known chaotic oscillators are: Lü that is based on saturated nonlinear function (SNLF) series and modeled by (1.2), where $f_0(x, k, m)$ is given in (1.3), with m being the slope, and Rössler that is modeled by (1.4). These and other integer-order chaotic systems have been implemented using commercially available electronic devices, integrated circuit technology [6], and using embedded systems like Arduino micro-controller [7], and field-programmable gate arrays (FPGAs), which are useful for also implementing different families of fractional-order chaotic oscillators [8]. Those chaotic circuits can be applied to generate random numbers [9], design chaotic secure communication systems [10], to control the movement of autonomous robots [11], and so on.

$$\dot{x} = y$$
$$\dot{y} = z \qquad (1.2)$$
$$\dot{z} = -ax - by - cz + d_1 f_0(x, k, m)$$

$$f_0(x, k, m) = \begin{cases} k, & \text{if} \quad x > 1, \\ mx, & \text{if} \quad |x| \leq 1, \\ -k, & \text{if} \quad x < 1 \end{cases} \qquad (1.3)$$

$$\dot{x} = -y - z$$
$$\dot{y} = x + ay \qquad (1.4)$$
$$\dot{z} = b + z(x - c)$$

1.2 Equilibrium Points and Eigenvalues

A chaotic oscillator must be analyzed to find their equilibrium points, which are used to evaluate the Jacobian matrix to find the eigenvalues that help determining the stability. This analysis is useful to estimate the step-size to solve the mathematical model of an integer-order chaotic oscillator and then to evaluate the Lyapunov exponents, fractal dimension, and entropy.

Let us consider the Lorenz oscillator given in (1.1), setting $\sigma = 10, \rho = 28$, and $\beta = 8/3$, the equilibrium points are found by setting to zero all state variables, thus leading to the equations in (1.5), from which the variables become: $x = y, z = \rho - 1$, and $y = \sqrt{\beta(\rho - 1)}$. This means that Lorenz chaotic oscillator has three equilibrium points: one located in the origin $EP_1 = (0, 0, 0)$ and the other two at $EP_2 = (\sqrt{\beta(\rho - 1)}, \sqrt{\beta(\rho - 1)}, \rho - 1)$ and $EP_3 = (-\sqrt{\beta(\rho - 1)}, -\sqrt{\beta(\rho - 1)}, \rho - 1)$. The Jacobian matrix of (1.1) at the equilibrium point $E^* = (x^*, y^*, x^*)$ is given by (1.6). From this Jacobian matrix, one can evaluate the characteristic equation of the form $|\lambda I - J_{E_1}| = 0$, leading to the system given in (1.7), from which one can find the eigenvalues of (1.1) for each equilibrium point. The resulting eigenvalues are shown in (1.8). A similar analysis is performed for other chaotic oscillators having self-excited chaotic attractors. Both the equilibrium points and eigenvalues can be used to identify the kind of dynamics, as already shown in [12].

$$
\begin{aligned}
\sigma(y - x) &= 0 \\
x(\rho - z) - y &= 0 \\
xy - \beta z &= 0
\end{aligned}
\tag{1.5}
$$

$$
J_{(x^*, y^*, z^*)} = \begin{pmatrix} \dfrac{\partial f_1}{\partial x} & \dfrac{\partial f_1}{\partial y} & \dfrac{\partial f_1}{\partial z} \\[2mm] \dfrac{\partial f_2}{\partial x} & \dfrac{\partial f_2}{\partial y} & \dfrac{\partial f_2}{\partial z} \\[2mm] \dfrac{\partial f_3}{\partial x} & \dfrac{\partial f_3}{\partial y} & \dfrac{\partial f_3}{\partial z} \end{pmatrix} = \begin{bmatrix} -\sigma & \sigma & 0 \\ \rho - z & -1 & -x \\ y & x & -\beta \end{bmatrix}
\tag{1.6}
$$

$$
\begin{bmatrix} \lambda & 0 & 0 \\ 0 & \lambda & 0 \\ 0 & 0 & \lambda \end{bmatrix} - \begin{bmatrix} -\sigma & \sigma & 0 \\ \rho - z & -1 & -x \\ y & x & -\beta \end{bmatrix} = \begin{bmatrix} \lambda + \sigma & -\sigma & 0 \\ -\rho & \lambda + 1 & x \\ -y & -x & \lambda + \beta \end{bmatrix}
\tag{1.7}
$$

$$
\begin{aligned}
E_1 &= \lambda_{(1,2,3)} = (-22.8277, 11.8277, -2.6677) \\
E_2 &= \lambda_{(1,2,3)} = (-13.8546, 0.0940, \pm 10.1945i) \\
E_3 &= \lambda_{(1,2,3)} = (-13.8546, 0.0940, \pm 10.1945i)
\end{aligned}
\tag{1.8}
$$

1.3 Numerical Methods for Integer-Order Chaotic Oscillators

Integer-order chaotic oscillators can be solved applying numerical methods, like the ones known as one-step and multistep. The mathematical models of the chaotic oscillators are described by ordinary differential equations (ODE) and nonlinear functions. These ODEs are discretized according to the numerical method that approaches the solution using computer resources.

Considering the ODE given in (1.9), its solution in the interval $a \le x \le b$, where a and b are finite, leads to an approximation not in the continuous interval but in the discrete set of points $x_n | n = 0, 1, \ldots, (b-a)/h$, and h is the step-size. Thus y_n is an approximation of the analytical solution of $y(x_n)$.

$$
\frac{dy}{dx} = f(x, y), \quad y(a) = \eta
\tag{1.9}
$$

The numerical methods for solving initial value problems like the ODEs modeling autonomous chaotic oscillators can be classified according to the way they compute the next value, depending just on past values or depending on past values and estimating values at the same iteration sub-index. They are classified as implicit and explicit methods. In both cases they also have different accuracy and convergence regions. Explicit methods compute y_{n+k}, using past values at iterations y_{n+j}, $f\{n+j\}$, $j = 0, 1, \ldots, k-1$. Implicit methods compute y_{n+k} updating values at the same iteration index (using an explicit method) and past values to approach the solution by $y_{n+k} = h\beta_k f(x_{n+k}, y_{n+k}) + g$, where β is a constant coefficient and g a function of values previously computed at y_{n+j}, $f\{n+j\}$, $j = 0, 1, \ldots, k-1$.

1.3.1 One-Step Methods: Forward Euler and Runge-Kutta

The simplest numerical method is known as Forward Euler and it is one-step because the solution at iteration $n+k$ depends just on one previous value, i.e., at one previous iteration. The approximation of the solution is based on the calculation of an estimated slope of a function that is extrapolated from an actual value to a next value. Due to its simplicity, this numerical method provides the lowest accuracy because the error compared to the analytical solution is very big if the time-step (h) is not the adequate. Forward-Euler is an explicit method and has the iterative equation shown in (1.10). On the side of the implicit numerical methods, and also one-step, the simplest one is known as Backward-Euler, which iterative equation is given in (1.11). It can be appreciated that the evaluation of y_{n+1} requires evaluating $f(x_{n+1}, y_{n+1})$, but it depends on $y_{n+1}!$, the value at the same iteration index $n + 1$. In this manner, this next value in $f(x_{n+1}, y_{n+1})$ is first computed by applying an explicit method like Forward Euler, and then one can approach the solution of ODEs combining implicit and explicit methods.

$$y_{n+1} = y_n + h \cdot f(x_n, y_n) \tag{1.10}$$

$$y_{n+1} = y_n + h \cdot f(x_{n+1}, y_{n+1}) \tag{1.11}$$

A one-step method that is quite accurate and very used in engineering for solving ODE systems is the fourth-order Runge-Kutta, which approaches the solution from Taylor series expansion. Table 1.1 shows the Runge-Kutta methods of orders first to fourth. These methods with orders higher than one are considered harder than a lineal multistep method, and because of the evaluations of middle points, they loss linearity. In addition, high-order Runge-Kutta methods require more hardware resources when they are implemented on embedded systems like FPGAs [13].

Table 1.1 Runge-Kutta numerical methods

Order	Iterative equation
First	$y_{j+1} = y_j + hf(y_j, t_j)$
Second	$y_{j+1} = y_j + hf\left(y_j + \frac{h}{2}f(y_j, t_j), t_k + \frac{h}{2}\right)$
Third	$y_{j+1} = y_j + \frac{h}{4}(k_1 + 3k_3)$ $k_1 = f(y_j, t_j)$ $k_2 = f\left(y_j + \frac{hk_1}{3}, t_j + \frac{h}{3}\right)$ $k_3 = f\left(y_j + \frac{2hk_2}{3}, t_j + \frac{2h}{3}\right)$
Fourth	$y_{j+1} = y_j + \frac{h}{6}(k_1 + 2k_2 + 2k_3 + k_4)$ $k_1 = f(y_j, t_j)$ $k_2 = f\left(y_j + \frac{hk_1}{2}, t_j + \frac{h}{2}\right)$ $k_3 = f\left(y_j + \frac{hk_2}{2}, t_j + \frac{h}{2}\right)$ $k_4 = f(y_j + hk_3, t_j + h)$

Table 1.2 Adams-Bashforth iterative equations for the first to the sixth orders

Order	Explicit multistep method
First	$y_{j+1} = y_j + hf(y_j, t_j)$
Second	$y_{j+1} = y_j + \frac{h}{2}\{3f(y_j, t_j) - f(y_{j-1}, t_{j-1})\}$
Third	$y_{j+1} = y_j + \frac{h}{12}\{23f(y_j, t_j) - 16f(y_{j-1}, t_{j-1}) + 5f(y_{j-2}, t_{j-2})\}$
Fourth	$y_{j+1} = y_j + \frac{h}{24}\{55f(y_j, t_j) - 59f(y_{j-1}, t_{j-1}) + 37f(y_{j-2}, t_{j-2})$ $-9f(y_{j-3}, t_{j-3})\}$
Fifth	$y_{j+1} = y_j + \frac{h}{720}\{1901f(y_j, t_j) - 2774f(y_{j-1}, t_{j-1}) + 2616f(y_{j-2}, t_{j-2})$ $-1274f(y_{j-3}, t_{j-3}) + 251f(y_{j-4}, t_{j-4})\}$
Sixth	$y_{j+1} = y_j + \frac{h}{1440}\{4277f(y_j, t_j) - 7923f(y_{j-1}, t_{j-1}) + 9982f(y_{j-2}, t_{j-2})$ $-7298f(y_{j-3}, t_{j-3}) + 2877f(y_{j-4}, t_{j-4}) - 475f(y_{j-5}, t_{j-5})\}$

1.3.2 Multistep Methods: Adams-Bashforth and Adams-Moulton

Different to the one-step methods, the multistep ones reuse past data, which time-steps define the number of steps. Multistep numerical methods are also known as linear multistep methods of step number n, or linear n-step methods. It means that an n-steps method requires n previous points $x_j, x_{j-1}, \ldots, x_{j-n+1}$, and also requires the value of the function f evaluated at those points to estimate the next value x_{j+1}. The general equation that describes the linear multistep methods is given in (1.12), where α_j and β_j are constants that take appropriate values to have different multistep methods [12, 14]. For instance, on the classification of explicit multistep methods one can find the Nth-order Adams-Bashforth, which are generated considering that $n = N$, $\beta_{-1} = 0$, and $\alpha_i = 0$ for $i = 1, 2, \ldots, N$. Table 1.2 shows these methods from the first to the sixth order.

$$y_{j+1} = \alpha_0 y_i + \alpha_1 y_{i-1} + \cdots + \alpha_{n-1} y_{j-n+1}$$

Table 1.3 Adams-Moulton iterative equations for the first to the sixth orders

Order	Implicit multistep method
First	$y_{j+1} = y_j + h f(y_{j+1}, t_{j+1})$
Second	$y_{j+1} = y_j + \frac{h}{2}\{f(y_{j+1}, t_{j+1}) + f(y_j, t_j)\}$
Third	$y_{j+1} = y_j + \frac{h}{12}\{5f(y_{j+1}, t_{j+1}) + 8f(y_j, t_j)) - f(y_{j-1}, t_{j-1})\}$
Fourth	$y_{j+1} = y_j + \frac{h}{24}\{9f(y_{j+1}, t_{j+1}) + 19f(y_j, t_j)) - 5f(y_{j-1}, t_{j-1})$ $+ f(y_{j-2}, t_{j-2})\}$
Fifth	$y_{j+1} = y_j + \frac{h}{720}\{251f(y_{j+1}, t_{j+1}) + 646f(y_j, t_j) - 264f(y_{j-1}, t_{j-1})$ $+106f(y_{j-2}, t_{j-2}) - 19f(y_{j-3}, t_{j-3})\}$
Sixth	$y_{j+1} = y_j + \frac{h}{1440}\{475f(y_{j+1}, t_{j+1}) + 1427f(y_j, t_j) - 798f(y_{j-1}, t_{j-1})$ $+482f(y_{j-2}, t_{j-2}) - 173f(y_{j-3}, t_{j-3}) + 27f(y_{j-4}, t_{j-4})\}$

$$+ h\left\{\beta_{-1}f(y_{j+1}, t_{j+1}) + \beta_0 f(y_j, t_j) + \beta_1 f(y_{j-1}, t_{j-1}) + \cdots \right.$$
$$\left. + \beta_{n-1}f(y_{j-n+1}, t_{j-n+1})\right\} \tag{1.12}$$

In the same manner as for the one-step methods, there exist the implicit multistep methods, which require an explicit method to evaluate the value required at the same iteration index. In this case, one well-known kind of implicit multistep methods are known as Nth-order Adams-Moulton, which are generated from (1.12) by setting $n = N - 1$ and $\alpha_i = 0$ for $i = 1, 2, \ldots, N - 1$. Table 1.3 shows these implicit methods from the first to the sixth orders.

1.3.3   Comparison of One-Step and Multistep Numerical Methods

Choosing the adequate numerical method for solving a given problem described by ODEs remains a challenge. In the same direction, the estimation of the time-step is a challenge to guarantee the lowest error with respect to an exact solution. That way, if one does not choose an appropriate numerical method and step-size, the approached solution may converge, diverge, be erroneous, or give rise to undesired computational effects like chaos or superstability [15, 16]. Table 1.4 shows the comparison of the global error $e_4 = |y(4) - y_4|$, which is generated when applying one-step (Forward Euler and fourth-order Runge-Kutta) and multistep (third-order Adams-Bashforth and second-order Adams-Moulton) methods, for the solution of the equation: $\frac{dy}{dt} = -y^2$, with $y(0) = 1$ for all cases. As one can see, the fourth-order Runge-Kutta provides the lowest error for this initial value problem, generating errors around 10^{-6} from $h = 2^{-1}$. This is a very high time-step value compared to the one using Forward-Euler, which reaches that error value but with $h = 2^{-14}$. This implies that the computing time is very heavy to reach the same value of time $t = 4$, at which the exact analytical solution becomes $y(4) = 0.2$. One

Table 1.4 Error comparison of several numerical integration methods for solving the initial value problem $dy = -y^2$ at time $t = 4$, with an analytical solution of $y(4) = 0.2$ and varying the step-sizes from 2^0 down to 2^{-14}

Step-size h	Forward-Euler		Third-order Adams-Bashforth		Second-order Adams-Moulton		Fourth-order Runge-Kutta	
	Approximate value y_4	Global error $e_4 = \lvert y(4) - y_4 \rvert$	Approximate value y_4	Global error $e_4 = \lvert y(4) - y_4 \rvert$	Approximate value y_4	Global error $e_4 = \lvert y(4) - y_4 \rvert$	Approximate value y_4	Global error $e_4 = \lvert y(4) - y_4 \rvert$
2^0	0.0	0.2	0.063328637116648	0.136671362883352	0.207187940988934	0.007187940988934	0.197676014825475	0.002323985174525
2^{-1}	0.1635534597328991	0.0364465402671009	0.180898176881417	0.019101823118583	0.204779729315604	0.004779729315604	0.200008068838967	8.068838967423275e-06
2^{-2}	0.183254883412050	0.016745116587950	0.197520889532756	0.002479110467244	0.201147711766030	0.001147711766030	0.200001848646586	1.848646585811409e-06
2^{-3}	0.191811518268007	0.008188481731993	0.199679695897706	3.203041022940001e-04	0.200268664858137	2.686648581370910e-04	0.200000130497778	1.304977784444272e-07
2^{-4}	0.195942986304455	0.004057013695545	0.1999589887738243	4.101226175701189e-05	0.200064849269855	6.484926984505570e-05	0.200000008350923	8.350923091926532e-09
2^{-5}	0.197980053169780	0.002019946830220	0.199994802941526	5.197058474015748e-06	0.200015918612094	1.591861209371270e-05	0.200000000524705	5.247049184031738e-10
2^{-8}	0.1997484400566650	2.515994333500138e-04	0.199999989719544	1.028045601425554e-08	0.200002447130045	2.447130448557378e-07	0.200000000000128	1.283417816466681e-13
2^{-10}	0.199937123550592	6.28764940800030e-05	0.199999999839143	1.608570221467431e-10	0.200000015267731	1.526773080184896e-08	0.200000000000000	2.775557561562891e-16
2^{-12}	0.199984282346833	1.571765316701890e-05	0.199999999997486	2.514016772536820e-12	0.200000000953813	9.538134726483349e-10	0.200000000000000	9.920072216264409e-16
2^{-14}	0.199996070677849	3.929321510139363e-06	0.199999999999963	3.699818229563334e-14	0.200000000059608	5.960748561406604e-11	0.200000000000000	1.110223024625157e-16

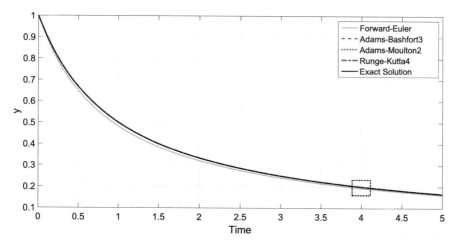

Fig. 1.2 Comparison of several numerical integration methods for the initial value problem $dy = -y^2$, approached with a step-size $h = 0.01$

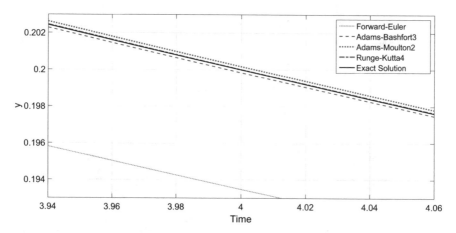

Fig. 1.3 Zoom from Fig. 1.2 to appreciate the errors generated by the one-step and multistep methods listed in Table 1.4

can also observe that both multistep methods approach the solution much better than Forward Euler, but not as better as when using the fourth-order Runge-Kutta.

The following figures show the comparison of the solutions for the initial value problem $dy = -y^2$, approached by one-step and multistep methods. Figure 1.2 shows the exact solution and the approaches by the four numerical methods that were compared in Table 1.4. In this case the four numerical methods were executed using $h = 0.01$. A better appreciation of the errors generated by each method is shown in Fig. 1.3, were a zoom is performed around the time $t = 4$. As mentioned above and from Table 1.4, Forward-Euler method generates the highest error, confirming that this is the most simple numerical method. The other methods generate a very similar error when compared to the exact solution, and the Fourth-order Runge-Kutta method almost equals the exact solution, so that a lot of researchers apply this method, which has low probability of generating undesired effects like computational chaos or superstability.

1.4   Numerical Simulation of Integer-Order Chaotic Oscillators

The simulation of integer-order and autonomous chaotic oscillators can be performed applying the one-step and multistep methods described above. For the case of the Lorenz chaotic oscillator [17], which is given in (1.1), to observe the phase-space portraits of the attractors for the different combinations of the state variables, the coefficient values are set to: $\sigma = 10$, $\rho = 28$, $\beta = 8/3$, and the initial conditions of the state variables (x, y, z) are $(0.1, 0.1, 0.1)$. In this manner, applying the four numerical methods given in Table 1.3, and using the same time-step of $h = 0.005$, Fig. 1.4 shows the phase-space portraits between two state variables of Lorenz chaotic attractor applying Forward Euler method, and the same characteristics are shown in Fig. 1.5 applying Adams-Bashforth, in Fig. 1.6 applying Adams-Moulton,

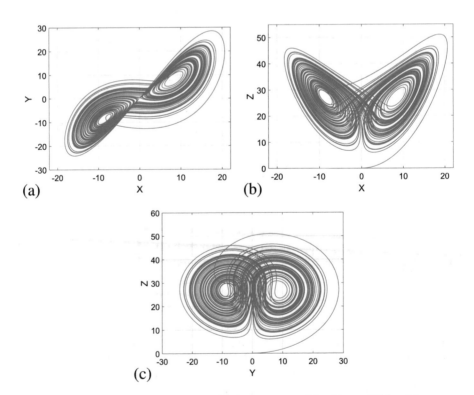

Fig. 1.4 Phase-space portraits of Lorenz chaotic attractor applying Forward Euler: (**a**) $x - y$, (**b**) $x - z$, and (**c**) $y - z$ portraits

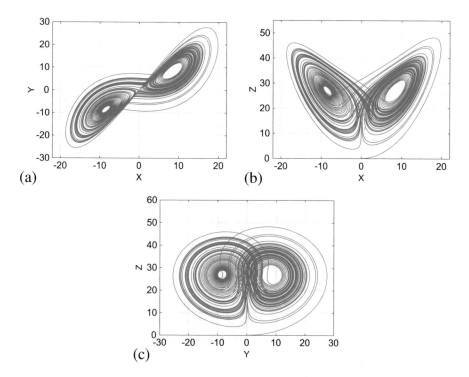

Fig. 1.5 Phase-space portraits of Lorenz chaotic attractor applying third-order Adams-Bashforth: (**a**) $x - y$, (**b**) $x - z$, and (**c**) $y - z$ portraits

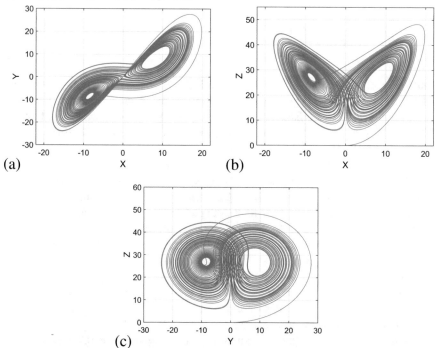

Fig. 1.6 Phase-space portraits of Lorenz chaotic attractor applying second-order Adams-Moulton: (**a**) $x - y$, (**b**) $x - z$, and (**c**) $y - z$ portraits

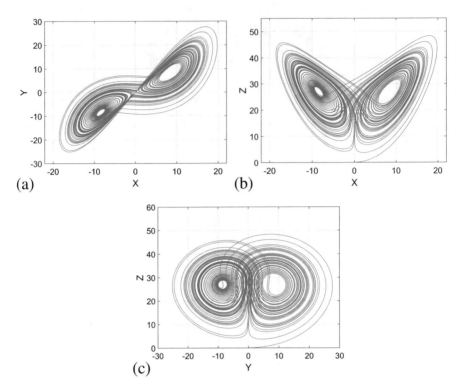

Fig. 1.7 Phase-space portraits of Lorenz chaotic attractor applying fourth-order Runge-Kutta: (**a**) $x - y$, (**b**) $x - z$, and (**c**) $y - z$ portraits

and in Fig. 1.7 applying fourth-order Runge-Kutta. As one sees, the ranges of values of the state variables are quite similar, as well as the chaotic attractors. However, their dynamical characteristics are a little bit different. This will be discussed further to show the evaluation of the Lyapunov exponents and Kaplan-Yorke dimension of several integer-order chaotic oscillators.

For the case of fractional-order chaotic oscillators, different numerical methods are applied, but the evaluation of the equilibrium points and eigenvalues is performed as shown in Sect. 1.2. In the same way, the following sections will show the evaluation of the dynamical characteristics of fractional-order chaotic oscillators, which can be designed from the integer-order ones, but performing the approximation to the solution of the fractional derivative of the state variables.

In the following figures, we show the phase-space portraits between two states variables of other integer-order chaotic oscillators, and applying the four numerical methods described above. Figure 1.8 shows the attractors of the Rössler chaotic oscillator [18] that are generated applying the four numerical methods again using $h = 0.005$. In this case, the equations modeling Rössler chaotic oscillator were given in (1.4), and the coefficient values are set to $a = 0.2, b = 0.2, c = 10$, with the initial conditions of the state variables (x, y, z) taken as $(0.5, 0.5, 0.5)$.

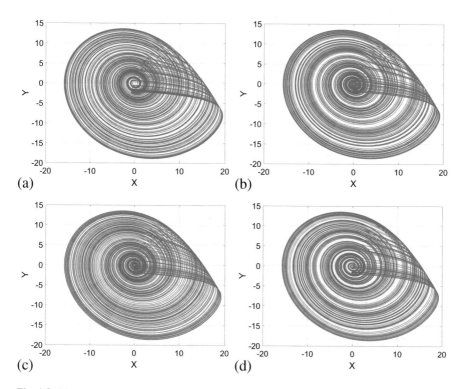

Fig. 1.8 Phase-space portraits of Rössler chaotic attractor between state variables $x - y$ and applying: (**a**) Forward Euler, (**b**) Adams-Bashforth, (**c**) Adams-Moulton, and (**d**) Runge-Kutta

The mathematical model of Chen's chaotic oscillator [19] is given in (1.13). It is simulated by setting their coefficients to $a = 35$, $b = 3$, and $c = 28$, and the initial conditions of the three state variables (x, y, z) equal to $(-3, -3, 15)$. Figure 1.9 shows the attractors in the $x - y$ phase-space portraits of the Chen chaotic oscillator that are generated applying the four numerical methods and with $h = 0.005$.

$$\begin{aligned}
\dot{x} &= a(y - x) \\
\dot{y} &= (c - a)x - xz + cy \\
\dot{z} &= xy - bz
\end{aligned} \tag{1.13}$$

Lü and Chen proposed a new integer-order chaotic oscillator [20], which is described by (1.14). The coefficient values to generate the chaotic attractor are set to: $a = 36$, $b = 3$, $c = 20$, and the simulation was performed by using the initial conditions of the three state variables (x, y, z) equal to $(0.2, 0.5, 0.3)$, and applying the four numerical methods described above. Figure 1.10 shows the attractors in the $x - y$ phase-space portraits that are generated with $h = 0.005$.

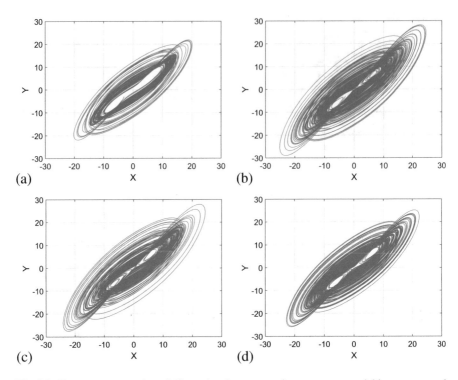

Fig. 1.9 Phase-space portraits of Chen chaotic attractor between state variables $x - y$ and applying: (**a**) Forward Euler, (**b**) Adams-Bashforth, (**c**) Adams-Moulton, and (**d**) Runge-Kutta

$$\begin{aligned}
\dot{x} &= a(y - x), \\
\dot{y} &= -xz + cy, \\
\dot{z} &= xy - bz
\end{aligned} \tag{1.14}$$

The chaotic oscillator introduced by Liu [21] is given in (1.15). The generation of chaotic attractors is done by setting the coefficient values to $a = 10, b = 40, k = 1, c = 2.5$, and $d = 4$. The phase-space portraits are generated with $h = 0.005$ and by using the initial conditions of (x, y, z) equal to (2.2, 2.4, 38). Figure 1.11 shows the attractors in the $x - y$ portraits.

$$\begin{aligned}
\dot{x} &= a(y - x), \\
\dot{y} &= bx - kxz, \\
\dot{z} &= -cz + dx^2
\end{aligned} \tag{1.15}$$

The Heart shape chaotic oscillator [22] is given in (1.16), where $G(z)$ is a nonlinear function that can be approached by a PWL function as shown in (1.17), and in which the parameters are set to $S_0 = 0.86, S_1 = 2.53, d_1 = 2$, and $d_2 = 8$. This chaotic oscillator is designed with the coefficient values equal to $m = 2$,

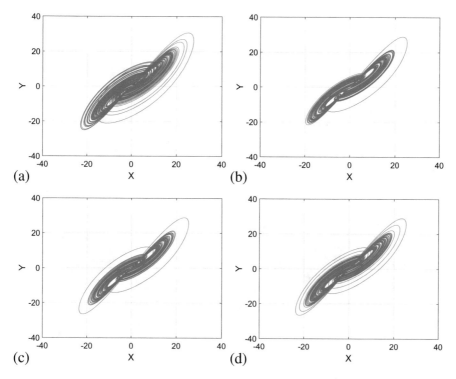

Fig. 1.10 Phase-space portraits of Lü-Chen chaotic attractor applying: (**a**) Forward Euler, (**b**) Adams-Bashforth, (**c**) Adams-Moulton, and (**d**) Runge-Kutta

$r = 0.5$, and the simulation is performed by using the initial conditions of the three state variables (x, y, z) equal to $(0.5, 1, 0.5)$. Figure 1.12 shows the attractors in the $x - y$ phase-space portraits that are generated with $h = 0.005$.

$$\dot{x} = y - x,$$
$$\dot{y} = sign(x)[1 - mz + G(z)], \qquad (1.16)$$
$$\dot{z} = |x| - rz$$

$$G(z) = \begin{cases} 0 & z < S_0 \\ d_1 & S_0 \leq z < S_1 \\ d_2 & z \geq S_1 \end{cases} \qquad (1.17)$$

Another kind of integer-order and autonomous chaotic oscillators are the ones proposed by Sprott [23], which are based in the multiplication of state variables. Sprott took as reference the Poincaré-Bendixson theorem [24], and then from (1.18) he discovered and selected the nineteen chaotic oscillators given in Table 1.5. These chaotic oscillators can also be simulated applying one-step and multistep numerical methods.

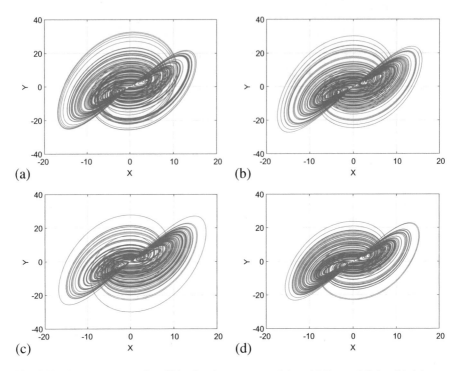

Fig. 1.11 Phase-space portraits of Liu chaotic attractor applying: (**a**) Forward Euler, (**b**) Adams-Bashforth, (**c**) Adams-Moulton, and (**d**) Runge-Kutta

The integer-order chaotic oscillators described above, namely: Lorenz described by (1.1), Lü described by (1.2), Rössler described by (1.4), Chen described by (1.13), Lü and Chen described by (1.14), Liu described by (1.15), Heart shape described by (1.16), the Sprott collection described by Table 1.5, and other integer-order chaotic oscillators can be simulated applying one-step and multistep methods. Other numerical methods can also be applied, which are more suitable for oscillatory systems, as shown in [25], where the authors apply the numerical method based on trigonometric polynomials, which outperforms Forward Euler and fourth-order RK, for instance. In that work, five integer-order chaotic oscillators are simulated and implemented on FPGAs, showing that trigonometric polynomials require lower hardware resources than Runge-Kutta.

All integer-order chaotic oscillators can be designed to behave as fractional-order ones. In that case, the derivatives become fractional, taken values between $0 < \alpha < 1$, where α is the order of the derivative. The special numerical methods applied to approach the solution of fractional-order chaotic oscillators are detailed in the following chapters, as well as their implementation details using amplifiers, FPAAs, and embedded systems like FPGAs.

$$\dot{x} = a + \sum_{i=1}^{3} b_i x_i + \sum_{i=1}^{3}\sum_{j=1}^{3} c_{i,j} x_i x_j \qquad (1.18)$$

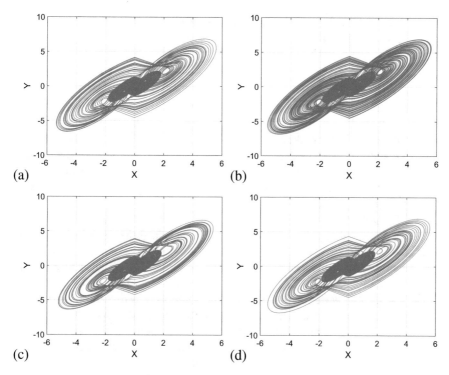

Fig. 1.12 Phase-space portraits of Heart shape chaotic attractor applying: (**a**) Forward Euler, (**b**) Adams-Bashforth, (**c**) Adams-Moulton, and (**d**) Runge-Kutta

Table 1.5 Nineteen autonomous chaotic oscillators proposed by Sprott [23]

Case	Equations		
A	$\dot{x} = y$	$\dot{y} = -x + yz$	$\dot{z} = 1 - y^2$
B	$\dot{x} = yz$	$\dot{y} = x - y$	$\dot{z} = 1 - xy$
C	$\dot{x} = yz$	$\dot{y} = x - y$	$\dot{z} = 1 - x^2$
D	$\dot{x} = -y$	$\dot{y} = x + z$	$\dot{z} = xz + 3y^2$
E	$\dot{x} = yz$	$\dot{y} = x^2 - y$	$\dot{z} = 1 - 4x$
F	$\dot{x} = y + z$	$\dot{y} = -x + \frac{1}{2}y$	$\dot{z} = x^2 - z$
G	$\dot{x} = 0.4x + z$	$\dot{y} = xz - y$	$\dot{z} = -x + y$
H	$\dot{x} = -y + z^2$	$\dot{y} = x + \frac{1}{2}y$	$\dot{z} = x - z$
I	$\dot{x} = -0.2y$	$\dot{y} = x + z$	$\dot{z} = x + y^2 - z$
J	$\dot{x} = 2z$	$\dot{y} = -2y + z$	$\dot{z} = -x + y + y^2$
K	$\dot{x} = xy - z$	$\dot{y} = x - y$	$\dot{z} = x + 0.3z$
L	$\dot{x} = y + 3.9z$	$\dot{y} = 0.9x^2 - y$	$\dot{z} = 1 - x$
M	$\dot{x} = -z$	$\dot{y} = x^2 - y$	$\dot{z} = 1.7 + 1.7x + y$
N	$\dot{x} = -2y$	$\dot{y} = x + z^2$	$\dot{z} = 1 + y - 2z$
O	$\dot{x} = y$	$\dot{y} = x - z$	$\dot{z} = x + xz + 2.7y$
P	$\dot{x} = 2.7y + z$	$\dot{y} = -x + y^2$	$\dot{z} = x + y$
Q	$\dot{x} = -z$	$\dot{y} = x - y$	$\dot{z} = 3.1x + y^2 + 0.5z$
R	$\dot{x} = 0.9 - y$	$\dot{y} = 0.4 + z$	$\dot{z} = xy - z$
S	$\dot{x} = -x - 4y$	$\dot{y} = x + z^2$	$\dot{z} = 1 + x$

1.5   Lyapunov Exponents, Kaplan-Yorke Dimension, and Entropy

Details on the analysis of the mathematical model of a chaotic oscillator and their numerical simulation can be found in [12], where one can see that a chaotic attractor can be a fixed point, a curve, a variety, or a strange attractor. An attractor is bounded and has a fractal dimension, in which its Hausdorff dimension can overcome its topological dimension [26]. The characteristics of a chaotic attractor can be quantified by evaluating its Lyapunov exponents, from which the Kaplan-Yorke dimension can be evaluated. That way, the majority of researchers agree that the main characterizations of chaotic systems are fractal dimension, Kolmogorov-Sinai entropy, and Lyapunov spectrum.

The Lyapunov exponents give the most characteristic description of the presence of a deterministic non-periodic flow [27], and they are taken as asymptotic measures characterizing the average rate of growth (or shrinking) of small perturbations to the solutions of a dynamical system. They also provide quantitative measures of response sensitivity of a dynamical system to small changes in initial conditions. That way, Lyapunov exponent values provide a means of ascertaining whether the behavior of a system is chaotic, so that the presence of at least one positive Lyapunov exponent in a dynamical system having at least three ODEs is often been taken as a signature of chaotic motion. In addition, a high value of the positive Lyapunov exponent indicates a high degree of unpredictability of the dynamical system, and therefore, the system has a more complex dynamic behavior.

Using the chaotic time series data that are generated after approaching the solutions applying numerical methods, one can use the time series analysis (TISEAN) program, already available on internet, to compute Lyapunov exponents, Kaplan-Yorke dimension and entropy, as well. On the other hand, beginning with the mathematical model of a chaotic oscillator, one can apply Wolf's method to determine the Lyapunov exponents [27], which for a third-order dynamical system, the three Lyapunov exponents take values as: positive, zero, and negative $(+,0,-)$. Applying numerical methods, the Lyapunov exponents are approached to be positive and negative, and the one that must be equal to zero can be approached by the one with the lowest value.

Varying the coefficients of a mathematical model, one can apply metaheuristics to maximize the positive or maximum Lyapunov exponent, as already shown in [28], where the case of study is the oscillator described by (1.2), which has four coefficients a, b, c, d_1 and they are varied in the range $[0.001 \ldots 1.000]$, in order to modify the standard settings of the coefficient values to maximize the positive Lyapunov exponent. One may also consider varying more than three significant digits, but in this particular case of varying three fractional values there exist 10^3 possible values for each coefficient, and, therefore, $(10^3)^4 = 10^{12}$ combinations for the complete dynamical system involving the four coefficients. That way, this huge search space justifies the application of metaheuristic optimization algorithms to maximize characteristics like positive Lyapunov exponent, fractal dimension, and entropy. It is worth mentioning that a slight variation in the design parameters of a chaotic oscillator, e.g., the coefficient values, may lead to a big change in the value of the Lyapunov exponents [27].

Finding the maximum value of the positive Lyapunov exponent becomes a difficult manual task. In this manner, works like [28, 29] apply metaheuristic algorithms to optimize the positive Lyapunov exponents and Kaplan-Yorke dimension, for instance. Those optimization algorithms can link either Wolf's method [27] or TISEAN to perform the evaluation of the dynamical characteristics. However, one can also link a procedure, as the following to evaluate the Lyapunov exponents of an ODE-based dynamical system.

Let us consider an n-dimensional dynamical system described as:

$$\dot{x} = f(x) \quad t > 0 \quad x(0) = x_0 \in \mathbb{R}^n \tag{1.19}$$

where x and f are n-dimensional vector fields. To determine the n-Lyapunov exponents of the dynamical system one has to find the long-term evolution of small perturbations to a trajectory, which are determined by the variational equation of (1.19), and given by (1.20), where J is the $n \times n$ Jacobian matrix of f. A solution to (1.20) with a given initial perturbation $y(0)$ can be written as (1.21), with $Y(t)$ as the fundamental solution satisfying (1.22), in which I_n denotes the $n \times n$ identity matrix.

$$\dot{y} = \frac{\partial f}{\partial x}\big(x(t)\big)y = J\big(x(t)\big)y \tag{1.20}$$

$$y(t) = Y(t)y(0) \tag{1.21}$$

$$\dot{Y} = J\big(x(t)\big)Y \quad Y(0) = I_n \tag{1.22}$$

If one considers the evolution of an infinitesimal n-parallelepiped $[p_1(t), \ldots, p_n(t)]$ with the axis $p_i(t) = Y(t)p_i(0)$, for $i = 1, \ldots, n$, where $p_i(0)$ denotes an orthogonal basis of $\mathbb{R}^n$, then the i-th Lyapunov exponent that measures the long-time sensitivity of the flow $x(t)$ with respect to the initial data $x(0)$ at the direction $p_i(t)$ is defined by the expansion rate of the length of the i-th axis $p_i(t)$ and is given by [27] (1.23).

$$\lambda_i = \lim_{t \to \infty} \frac{1}{t} \ln \|p_i(t)\| \tag{1.23}$$

In general, the method used in [28] can be summarized as follows:

1. The initial conditions of the original chaotic dynamical system and its variational system are set to $\mathbf{X_0}$ and $\mathbf{I_{n \times n}}$, respectively.
2. The original and variational systems are integrated along several time steps until an orthonormalization period TO is reached. The integration of the variational

system $\mathbf{Y} = [y_1, y_2, y_3]$ depends on the specific Jacobian that the original chaotic dynamical system $\mathbf{X}$ is using in the current time-step h.

3. The variational system is orthonormalized applying the standard Gram-Schmidt method, and the logarithm of the norm of each Lyapunov vector contained in $\mathbf{Y}$ is obtained and accumulated in time.

4. The next integration is carried out by using the new orthonormalized vectors as initial conditions. This process is repeated until the full integration period T is reached.

5. Therefore, the Lyapunov exponents are obtained by evaluating (1.24), in which the time-step selection can be done by using the minimum absolute value of all the eigenvalues of the system λ_{min}, and ψ can be chosen well above the sample theorem as 50, to set $t_{step} = \frac{1}{\lambda_{min}\psi}$

$$\lambda_i \approx \frac{1}{\mathbf{T}} \sum_{j=\mathrm{TO}}^{\mathbf{T}} \ln \|\mathbf{y_i}\| \tag{1.24}$$

The authors in [29] introduced an approach to optimize Kaplan-Yorke dimension (D_{KY}) in integer-order chaotic oscillators applying metaheuristics because when varying the coefficients of the mathematical model, a very huge number of combinations arise and the required computing time can be unreachable. Wolf's method [27] was applied to evaluate Lyapunov exponents and D_{KY}, and the authors sketched the implementation of a chaotic secure communication system to highlight that the best encryption of an image can be performed by the chaotic oscillator having high values of D_{KY}.

For instance, Rössler system given in (1.4) was used as case of study by setting $a = 0.15, b = 0.20$, and $c = 10$, which provided a positive Lyapunov exponent LE+ = 0.13 and $D_{KY} = 2.01$. Also, the Lorenz chaotic system given in (1.1) was optimized with its design parameters: $\sigma = 0.16, \rho = 45.92$, and $\beta = 4$, for which the numerical simulation provided LE+ = 2.16 and $D_{KY} = 2.07$. In [29], differential evolution and particle swarm optimization algorithms were applied to maximize D_{KY}. The cases of study were chaotic oscillators modeled by three ODEs thus providing three Lyapunov exponents $(+,0,-)$, which were used to evaluate the Kaplan-Yorke dimension using (1.25), where k is an integer such that the sum of the Lyapunov exponents (λ_i) is nonnegative. If chaotic behavior is guaranteed in (1.1) and (1.4), then $k = 2$, so that λ_{k+1} is the third Lyapunov exponent, and the dimension D_{KY} is higher than 2.

$$D_{KY} = k + \frac{\sum_{i=1}^{k} \lambda_i}{\lambda_{k+1}} \tag{1.25}$$

The equilibrium points are evaluated as shown in Sect. 1.2, so that the associated Jacobians of (1.1) and (1.4) are shown in Table 1.6, so that each chaotic oscillator has three eigenvalues for each equilibrium point. For more complex systems the eigenvalues can be calculated applying Cardano's method [30]. The initial

Table 1.6 Jacobian and equilibrium points of the chaotic systems given in (1.1) and (1.4)

System	Jacobian	Equilibrium points
Lorenz	$\begin{bmatrix} -\sigma & \sigma & 0 \\ \rho - z & -1 & -x \\ y & x & \beta \end{bmatrix}$	$\left(\pm\sqrt{\beta(\rho - 1)}, \pm\sqrt{\beta(\rho - 1)}, \rho - 1 \right)$
Rössler	$\begin{bmatrix} 0 & -1 & -1 \\ 1 & a & 0 \\ z & 0 & x - c \end{bmatrix}$	$\left(\dfrac{c \pm \sqrt{c^2 - 4ab}}{2}, -\dfrac{c \pm \sqrt{c^2 - 4ab}}{2a}, \dfrac{c \pm \sqrt{c^2 - 4ab}}{2a} \right)$

Table 1.7 Results of 10 runs performed by DE and PSO algorithms in [29] for (1.1) and (1.4), showing $D_{KY} > 2$

Oscillator	Runs	DE			PSO		
		Max D_{KY}	Mean	σ	Max D_{KY}	Mean	σ
Lorenz	1	2.0754	2.0773	0.01486	2.0719	2.0713	0.00969
	2	2.0730	2.0732	0.01481	2.0718	2.0712	0.00969
	3	2.0843	2.0783	0.01466	2.0706	2.0703	0.00969
	4	2.0791	2.0775	0.01478	2.0712	2.0705	0.00969
	5	2.0785	2.0772	0.01474	2.0712	2.0702	0.00969
	6	2.0839	2.0770	0.01486	2.0662	2.0710	0.00969
	7	2.0796	2.0752	0.01487	2.0690	2.0702	0.00969
	8	2.0739	2.0738	0.01479	2.0684	2.0708	0.00969
	9	2.0733	2.0733	0.01451	2.0692	2.0703	0.00969
	10	2.0741	2.0740	0.01431	2.0714	2.0744	0.00969
Rössler	1	2.0250	2.0395	0.02036	2.0235	2.0230	0.03285
	2	2.0350	2.0377	0.02043	2.0225	2.0224	0.03286
	3	2.0510	2.0364	0.02046	2.0270	2.0210	0.03286
	4	2.0300	2.0374	0.02048	2.0229	2.0240	0.03289
	5	2.0500	2.0381	0.02038	2.0198	2.0220	0.03285
	6	2.0203	2.0383	0.02039	2.0269	2.0240	0.03288
	7	2.0203	2.0363	0.02028	2.0233	2.0230	0.03288
	8	2.0204	2.0373	0.02044	2.0170	2.0240	0.03287
	9	2.0204	2.0405	0.02026	2.0254	2.0230	0.03285
	10	2.0055	2.0414	0.02039	2.0265	2.0220	0.03287

conditions (x_0, y_0, z_0) were established to be (0.1, 0.1, 0.1) for Lorenz and (0.5, 0.5, 0.5) for Rössler.

The DE and PSO algorithms were executed in [29], with the same number of populations (P) and maximum generations (G), to maximize D_{KY}, which results are given in Table 1.7. As one can see, $D_{KY} > 2$ in all cases and the maximum variation is approximately ± 0.05. The authors demonstrated through the encryption of an image that the highest the value of D_{KY}, the better encrypted image.

1.6   Fractional-Order Chaotic Systems

Although nowadays one can find many applications of autonomous integer-order chaotic oscillators, the recent research on fractional-order ones promises enhancing those applications. The majority of researchers agree that the main reason for the delay on dealing with fractional-order chaotic oscillators was the absence of solution methods for fractional-order differential equations. Therefore, and as already mentioned in [31], at the present time there are many methods for approximation of the fractional derivative and integral, and fractional calculus can easily be used in wide areas of applications like in physics, electrical engineering, control systems, robotics, signal processing, chemical mixing, bioengineering, and so on. All reviews on the history of fractional calculus agree that this topic began more than 300 years ago, and the first reference may probably being associated with Leibniz and L'Hospital in 1695, where the half-order derivative was mentioned.

Recently, various mathematical definitions of chaos have been introduced, and all of them are based on the characteristic related to the super-sensitivity or sensitive dependence on the initial conditions, which are characterized by Lyapunov instability as a main property of the chaotic oscillation. In [31] one can find more information on the fundamentals of the theory of nonlinear oscillations, which were laid in 1960s and 1970s by A. Poincaré, B. Van der Pol, A. A. Andronov, N. M. Krylov, A. N. Kolmogorov, D. V. Anosov, Ya. G. Sinai, V. K. Mel'nikov, Yu. I. Neimark, L. P. Shil'nikov, G. M. Zaslavsky, and their collaborators. In that book the authors provided a study of fractional-order chaotic systems accompanied by MatLab programs for simulating their state space trajectories.

One of the seminal papers dealing with fractional-order circuits is [32], which discusses the major disadvantages of integer-order sinusoidal oscillators, and then the authors proposed using noninteger-order filters to make oscillators of fractional order greater than 2. That paper showed that the application of a criterion globally characterizing the performances of systems just above the threshold of oscillation as regards the purity of the signal and the stability of the frequency, enabled to determine an optimal fractional order $n = 2.458$. Afterwards, an experimental arrangement of order 5/2 produced a modulation characteristic with displacement error less than 5% for values of the relative deviation of frequency between -50 and $+50\%$. More recently, a book on nonlinear noninteger-order circuits and systems was published in 2002 [33], which includes a theoretical introduction to noninteger-order systems, as well as several applications in control, noninteger-order cellular neural networks, and other circuit realizations of noninteger-order systems. Numerical methods for solving autonomous fractional-order chaotic systems were also introduced. For instance, linear transfer function approximations of the fractional integrator block were calculated in [34], for a set of fractional orders in (0,1], based on frequency domain arguments, and the resulting equivalent models were studied. Two chaotic models were considered: an electronic chaotic oscillator and a mechanical chaotic jerk model. The linear transfer function approximations showed their usefulness in simulating different types of model nonlinearities, and using the proper control parameters, chaotic attractors were obtained with system orders as low as 2.1. On this direction, some recent contributions on the optimization, control, circuit realizations, and applications of fractional-order systems were collected in [35], highlighting multidisciplinary applications, and implementations on FPGA, integrated circuits, memristors, control algorithms, photovoltaic systems, robot manipulators, oscillators, etc. The applications were also given in both continuous-time and discrete-time dynamics and chaotic systems. To provide enhanced applications, the authors in [36] stated that discovering new chaotic systems with interesting dynamical features will be of great interest in the coming years. One such important and interesting feature is the type and shape of equilibrium points, which improve an application on chaos-based secure communications by the differential chaos shift keying method. Another recent and important line for research is highlighted in [37], which introduces the novel classification of nonlinear dynamical systems including two kinds of attractors: self-excited attractors and hidden attractors. Those authors demonstrate that the localization of self-excited attractors by applying a standard computational procedure is straightforward, but hidden attractors are difficult to locate thus requiring specific computational procedures because the equilibrium points do not help in the localization of hidden attractors. Examples of chaotic dynamical systems embedding hidden attractors are systems with no equilibrium points, with only stable equilibria, curves of equilibria, and surfaces of equilibria, and with non-hyperbolic equilibria.

On the practical aspects, the authors in [38] discuss how fractional dynamics and control can be used to solve nonlinear science and complexity issues in the areas of nonlinear dynamics, vibration and control with analytical, numerical, and experimental results. That work describes new methods for control and synchronization of fractional-order dynamical systems. Generally, the majority of papers introduce master-slave synchronization techniques of fractional-order chaotic systems [39]. Some books like [40] address the need to couple a definition of the fractional differentiation or integration operator with the types of dynamical systems that are analyzed, in which the focus is on basic aspects of fractional calculus, such as stability analysis that is required to tackle synchronization in coupled fractional-order systems. The authors of the book in [41] reported the latest advances and applications of fractional-order control and synchronization of chaotic systems, addressing theories, methods, and applications in a number of research areas related to fractional-order control and synchronization of chaotic systems, such as fractional-order chaotic systems, hyperchaotic systems, complex systems, fractional-order discrete chaotic systems, chaos control, chaos synchronization, jerk circuits, fractional chaotic systems with hidden attractors, neural network, fuzzy logic controllers, behavioral modeling, robust and adaptive control, sliding mode control, different types of synchronization, and circuit realization of chaotic systems. The authors of the book in [42] also reported outstanding research on model-based control design methods for systems described by fractional dynamic models.

As mentioned in the previous sections, numerical methods for fractional-order chaotic systems are quite different than those applied to solve integer-order ones.

For instance, some numerical simulation approaches of fractional-order systems can be found in [31], and more recently, some time-domain numerical methods have been studied and implemented on FPGAs, such as Grünwald-Letnikov method in [8], and Adam-Bashforth-Moulton method in [43]. On another direction, several implementations and applications of fractional-order dynamical systems have been presented for the simulation and FPGA-based implementation of the control for a class of variable-order fractional chaotic systems by using the sliding mode control strategy [44]. Other applications are for example: the dynamical analysis and chaos control in a fractional-order synchronous reluctance motor [45]; the control of hidden chaos applying nonlinear feedback controllers, sliding mode controllers and hybrid combination of them [46]; the generation of a four-wing chaotic attractor [47]; the control of a 4-D nonautonomous fractional-order uncertain model of a proportional-integral (PI) speed-regulated current-driven induction motor using a fractional-order adaptive sliding mode controller [48]; the chaotic oscillations in a fractional-order model of a portal frame with nonideal loading [49]; the analysis of different chaotic systems on the basis of hardware resource utilization, static power analysis, and synthesis frequency on FPGA [50]; the chaos control in fractional-order smart grid with adaptive sliding mode control and genetically optimized PI-derivative (PID) control [51]; and so on.

The year 1695 was the year of the introduction of fractional calculus that was developed as pure mathematics area [52]. Nowadays, several researchers have demonstrated that fractional derivatives are extremely useful to model the behavior of all real behaviors in all fields like in acoustics, thermal systems, materials and mechanical systems, signal processing, systems identification, reconfigurable hardware, etc. [53]. Those applications exploit the advantages of using fractional-order systems against integer ones. Fractional calculus is useful in different engineering and science areas [54]; physics, biology, chemistry, geology, control theory, electromagnetism [55], electrical circuits, image processing, optics [56], nanotechnology [57], secure communications, and so on. Some applications apply synchronization techniques to fractional-order chaotic oscillators, for example in diagnosing in an intelligent way solar energy networks [58], or as in [59] for image encryption, in [60] to develop the fractional-order models for the infection of Ebola virus in heterogenous complex networks, or in [61] to enhance image encryption using a fractional-order relaxation oscillator and the model Quasi Gamma Curve (QGC).

1.7 Definitions of Fractional-Order Derivatives and Integrals

Fractional calculus is a generalization of integration and differentiation to the noninteger-order fundamental operator $_aD_a^t$, where a and t are the bounds of the operation and $\alpha \in R$. The continuous integro-differential operator is defined as

$$
{}_\alpha D_a^t = \begin{cases} \frac{d^\alpha}{dt^\alpha}, & \alpha > 0 \\ 1, & \alpha = 0 \\ \int_a^t (d\tau)^\alpha, & \alpha < 0 \end{cases} \tag{1.26}
$$

The three most frequently used definitions for the general fractional differintegral are: Grünwald-Letnikov, Riemann-Liouville, and Caputo definitions [52, 62, 63]. Other definitions are connected with well-known names as, for instance, Weyl, Fourier, Cauchy, Abel, Nishimoto, etc. In this book, Grünwald -Letnikov, Riemann-Liouville, and Caputo definitions are considered. This consideration is based on the fact that for a wide class of functions, these three best known definitions are equivalent under some conditions [63].

1.7.1   Grünwald-Letnikov Fractional Integrals and Derivatives

Let us consider the continuous function $f(t)$. Its first derivative can be expressed as (1.27), where when this is used twice, one obtains a second derivative of the function $f(t)$ in the form of (1.28). Afterwards, using both (1.27) and (1.28) one can get a third derivative of the function $f(t)$ in the form of (1.29).

$$
\frac{d}{dt} f(t) \equiv f'(t) = \lim_{h \to 0} \frac{f(t) - f(t-h)}{h} \tag{1.27}
$$

$$
\begin{aligned}
\frac{d^2}{dt^2} f(t) \equiv f''(t) &= \lim_{h \to 0} \frac{f'(t) - f'(t-h)}{h} \\
&= \lim_{h \to 0} \frac{1}{h} \left\{ \frac{f(t)-f(t-h)}{h} - \frac{f(t-h)-f(t-2h)}{h} \right\} \\
&= \lim_{h \to 0} \frac{f(t) - 2f(t-h) + f(t-2h)}{h^2}
\end{aligned} \tag{1.28}
$$

$$
\frac{d^3}{dt^3} \equiv f'''(t) = \lim_{h \to 0} \frac{f(t) - 3f(t-h) + 3f(t-2h) - f(t-3h)}{h^3} \tag{1.29}
$$

According to this rule, one can write a general formulae for the n-th derivative of a function $f(t)$, by t for $n \in N$, $j > n$ in the form of (1.30), which expresses a linear combination of function values $f(t)$ in the variable t. Henceforth, Binomial coefficients with alternating signs for positive values of n are defined by (1.31), while in the case of negative values of n, one gets (1.32), where $\begin{bmatrix} n \\ j \end{bmatrix}$ is defined by (1.33).

$$\frac{d^n}{dt^n} f(t) \equiv f^{(n)}(t) = \lim_{h \to 0} \frac{1}{h^n} \sum_{j=0}^{n} (-1)^j \binom{n}{j} f(t - jh) \tag{1.30}$$

$$\binom{n}{j} = \frac{n(n-1)(n-2)\cdots(n-j+1)}{j!} = \frac{n!}{j!(n-j)!} \tag{1.31}$$

$$\binom{-n}{j} = \frac{-n(-n-1)(-n-2)\cdots(-n-j+1)}{j!} = (-1)^j \begin{bmatrix} n \\ j \end{bmatrix} \tag{1.32}$$

$$\begin{bmatrix} n \\ j \end{bmatrix} = \frac{n(n+1)\cdots(n+j-1)}{j!} \tag{1.33}$$

Substituting $-n$ in (1.30) by n, one can write (1.34), where n is a positive integer number.

$$\frac{d^{-n}}{dt^{-n}} f(t) \equiv f^{(-n)}(t) = \lim_{h \to 0} \frac{1}{h^n} \sum_{j=0}^{n} \begin{bmatrix} n \\ j \end{bmatrix} f(t - jh) \tag{1.34}$$

According to (1.27)–(1.30), one can write the fractional-order derivative definition of order α, ($\alpha \in R$) by t, which has the form of (1.35). In this case, the binomial coefficients can be calculated by using the relation between Euler's Gamma function and factorial, which becomes defined as in (1.36), for $\binom{a}{0} = 1$.

$$D_t^\alpha f(t) = \lim_{h \to 0} \frac{1}{h^\alpha} \sum_{j=0}^{\infty} (-1)^j \binom{a}{j} f(t - jh) \tag{1.35}$$

$$\binom{a}{j} = \frac{\alpha!}{j!(\alpha - j)!} = \frac{\Gamma(\alpha + 1)}{\Gamma(j + 1)\Gamma(\alpha - j + 1)} \tag{1.36}$$

The definition given in (1.36) requires the sum of infinity data, which is not reachable using power computing resources. In this manner, one can consider $n = \frac{t-a}{h}$, and supposing that α is a real constant, which expresses a limit value, then (1.36) can be updated to the equation given in (1.37), where $[x]$ is associated with the integer part of x, and α and t are the bounds of the operation in $_a D_t^\alpha f(t)$. These equations will be discussed in Chap. 5 for the FPGA-based implementation of fractional-order chaotic oscillators.

$$_a D_t^\alpha f(t) = \lim_{h \to 0} \frac{1}{h^\alpha} \sum_{j=0}^{\left[\frac{t-a}{h}\right]} (-1)^j \binom{a}{j} f(t - jh) \tag{1.37}$$

1.7.2 Riemann-Liouville Fractional Integrals and Derivatives

The Riemann-Liouville definition can be described under some assumptions, for example one can consider the Riemann-Liouville n-fold integral, which can be defined by (1.38) for $n \in N, n > 0$.

$$\underbrace{\int_a^t \int_a^{t_n} \int_a^{t_{n-1}} \cdots \int_a^{t_3} \int_a^{t_2} f(t_1) dt_1 dt_2 \cdots dt_{n-1} dt_n}_{n\text{-fold}} = \frac{1}{\Gamma(n)} \int_a^t \frac{f(\tau)}{(t-\tau)^{1-n}} d\tau$$

(1.38)

The fractional-order integral α for the function $f(t)$ can then be expressed from (1.38) as given in (1.39), for $\alpha, a \in R, \alpha < 0$.

$$_a I_t^\alpha f(t) \equiv {}_a D_t^{-\alpha} f(t) = \frac{1}{\Gamma(-\alpha)} \int_a^t \frac{f(\tau)}{(t-\tau)^{\alpha+1}} d\tau$$

(1.39)

From the relation expressed in (1.39), one can write the formulae for the Riemann-Liouville definition of the fractional derivative of order α in the form of (1.40), for $(n-1 < \alpha < n)$, where α and t are the limits of the operator $_a D_t^\alpha f(t)$.

$$_a D_t^\alpha f(t) = \frac{1}{\Gamma(n-\alpha)} \frac{d^n}{dt^n} \int_a^t \frac{f(\tau)}{(t-\tau)^{\alpha-n+1}} d\tau$$

(1.40)

For the particular case of considering that $0 < \alpha < 1$ and $f(t)$ being a causal function of t, that is, $f(t) = 0$ for $t < 0$, then the fractional-order integral is defined by (1.41), and the expression for the fractional-order derivative is defined by (1.42), where $\Gamma(\cdot)$ is the Euler's Gamma function [52].

$$_0 D_t^{-\alpha} f(t) = \frac{1}{\Gamma(\alpha)} \int_0^t \frac{f(\tau)}{(t-\tau)^{1-\alpha}} d\tau, \, for \, 0 < \alpha < 1, t > 0$$

(1.41)

$$_0 D_t^\alpha f(t) = \frac{1}{\Gamma(n-\alpha)} \frac{d^n}{dt^n} \int_0^t \frac{f(\tau)}{(t-\tau)^{\alpha-n+1}} d\tau$$

(1.42)

1.7.3 Caputo Fractional Derivatives

The Caputo definition of fractional-order derivatives can be written as given in (1.43) [63, 64]. As considered in several works, under the homogenous initial conditions the Riemann-Liouville and the Caputo derivatives are equivalent. For instance, describing the Riemann-Liouville fractional derivative by $_a^{RL} D_t^a t(t)$, and the Caputo definition by $_a^C D_t^a f(t)$, then the equivalent relationship between them is given in (1.44), for $f^{(k)}(a) = 0$, and with $(k = 0, 1, \ldots, n-1)$.

$$_a D_t^\alpha f(t) = \frac{1}{\Gamma(n-\alpha)} \int_a^t \frac{f^{(n)}(\tau)}{(t-\tau)^{\alpha-n+1}} d\tau, \; for \; n-1 < \alpha < n \qquad (1.43)$$

$$_a^{RL} D_t^a t(t) = {}_a^C D_t^a f(t) + \sum_{k=0}^{n-1} \frac{(t-a)^{k-a}}{\Gamma(k-\alpha+1)} f^{(k)}(a) \qquad (1.44)$$

The initial conditions for the fractional-order differential equations with the Caputo derivatives are in the same form as for the integer-order differential equations. This is an advantage because the majority of problems require definitions of fractional derivatives, where there are clear interpretations of the initial conditions, which contain $f(a)$, $f'(a)$, $f''(a)$, etc.

1.8 Numerical Methods for Fractional-Order Chaotic Oscillators

The fractional-order derivatives and integrals are generalizations of the integer-order ones that are particular cases. In addition, it is worth mentioning that any dynamical system of integer-order can be adapted to a system of arbitrary order, including fractional-order. However, yet there is no generalized consensus on the definition of the fractional-order derivatives and integrals, so that one can find a variety of proposals; among them the most known and applied in different research areas are fractional derivative and integral of Riemmann-Liouville, Grünwald-Letnikov, and Caputo [65].

The approximation of the solution of the fractional-order derivatives can be performed applying the definitions of Riemmann-Liouville, Grünwald-Letnikov, and Caputo. In this book, the time domain approximations well known as Grünwald-Letnikov, which approximation expression in given in (1.45), and the predictor-corrector Adams-Bashforth-Moulton method given in (1.46) and (1.47), respectively, are used to simulate fractional-order oscillators that are implemented using amplifiers, FPAAs, and FPGAs in the following chapters.

$$_a D_t^\alpha f(t) = \lim_{h \to 0} \frac{1}{h^q} \sum_{j=0}^{[\frac{t-a}{h}]} (-1)^j \binom{q}{j} f(t-jh) \qquad (1.45)$$

$$y_h^p(t_n+1) = \sum_{k=0}^{m-1} \frac{t_{n+1}^k}{k} y_0^{(k!)} + \frac{1}{\Gamma(q)} \sum_{k=0}^{n} b_{j,n+1} f(t_j, y_n(t_j)) \qquad (1.46)$$

$$y_h(t_{n+1}) = \sum_{k=0}^{m-1} \frac{t_{n+1}^k}{k!} y_0^{(k)} + \frac{h^q}{\Gamma(\alpha+2)} f\left(t_{n+1}, y_h^p(t_{n+1})\right)$$

$$+ \frac{h^q}{\Gamma(\alpha+2)} \sum_{j=0}^{n} a_{j,n+1} f(t_j, y_n(t_j)). \tag{1.47}$$

The numerical solution of the fractional-order derivatives, as those ones modeling fractional-order chaotic oscillators, is approached herein applying the definitions associated with the Grünwald-Letnikov and Adams-Bashforth-Moulton methods. In several articles, one can find that for a wide class of functions, both definitions are equivalent. For instance, the relation of Grünwald-Letnikov with the explicit approximation given in (1.46) for the q-th derivative at the points kh, $(k = 1, 2, \dots)$ has the form of (1.48) [63, 65, 66], where L_m denotes the "memory length," $t_k = kh$, h is related to the time-step h, and $(-1)^j \binom{q}{j}$ are the binomial coefficients $C_j^q (j = 0, 1, \dots)$. In general, these binomial coefficients can be approached by an infinite number of values; however, it will be very difficult to implement on digital hardware. In this manner, one can take advantage of the concept on "short memory" [65], which is commonly applied in the Grünwald-Letnikov method to reduce hardware resources, and which memory length depends on the behavior of the coefficients given in (1.49).

$$_{(k-L_m/h)}D_{tk}^q f(t) \approx h^{-q} \sum_{j=0}^{k} (-1)^j \binom{q}{j} f\left(t_{k-j}\right) \tag{1.48}$$

$$c_0^q = 1, \qquad c_j^q = \left(1 - \frac{1+q}{j}\right) c_{j-1}^q \tag{1.49}$$

Applying the Grünwald-Letnikov approximation of the fractional-order derivative of the form: $_aD_t^q y(t) = f(y(t), t)$, leads us to deal with the discrete equation given in (1.50), which takes advantage of the short memory concept and then the binomial coefficients are truncated by a desired length of memory, which is suitable for hardware implementation like on an FPGA.

$$y(t_k) = f(y(t_k), t_k)h^q - \sum_{j=v}^{k} c_j^{(q)} y(t_{k-j}) \tag{1.50}$$

The sum in (1.50) is associated with the memory of the algorithm, and it can be truncated according to the short memory principle [65]. That way, the lowest index in the sum operation should be $v = 1$ for the case when $k < (L_m/h)$, and $v = k - (L_m/h)$ for the case $k > (L_m/h)$. Besides, without applying the short memory principle one must set $v = 1$ for all values of k. It is then quite obvious that when performing this truncation the error should increase and then the solution may not converge. Besides, one can estimate the length of memory, for example: if $f(t) \leqslant M$, L_m can be estimated using (1.51), which barely includes a required precision ε.

$$L \geqslant \left(\frac{M}{\varepsilon |\Gamma(1 - q)|} \right)^{1/q} \tag{1.51}$$

The time domain Grünwald-Letnikov method is an explicit one, and as mentioned above, one can apply an implicit one, for instance the well-known predictor-corrector Adams-Bashforth-Moulton method [67]. This approximation for the fractional-order derivatives of the chaotic oscillators is more exact than Grünwald-Letnikov, but it requires higher number of operations and hardware resources. The Adams-Bashforth-Moulton method is based on the fact that the fractional-order derivative of the form given in (1.52) is equivalent to Volterra's integral equation given in (1.53).

$$D_t^q y(t) = f(y(t), t), \; y^{(k)}(0) = y_0^{(k)}, \, k = 0, 1, \dots, m - 1 \tag{1.52}$$

$$y(t) = \sum_{k=0}^{\lceil q \rceil - 1} y_0^{(k)} \frac{t^k}{k!} + \frac{1}{\Gamma(q)} \int_0^t (t - \tau)^{q-1} f(\tau, y(\tau)) d\tau \tag{1.53}$$

Discretizing (1.53) for a uniform array $t_n = nh$ ($n = 0, 1, \dots, N$), $h = T_{sim}/N$ and using the short memory principle (fixed or logarithmic [68]) one gets a numerical approximation very close to the truth solution of $y(t_n)$ of the fractional differential equation, while conserving the exactness of the fractional-order. Therefore, supposing that one has the approximations in $y_h(t_j)$, $j = 1, 2, \dots, n$, then one can get the solution to $y_h(t_{n+1})$ using (1.54), where $a_{j,n+1}$ is evaluated by (1.55).

$$y_h(t_{n+1}) = \sum_{k=}^{m-1} \frac{t_{n+1}^k}{k!} y_0^{(k)} + \frac{h^q}{\Gamma(\alpha + 2)} f(t_{n+1}, y_h^p(t_{n+1}))$$

$$+ \frac{h^q}{\Gamma(\alpha + 2)} \sum_{j=0}^{n} a_{j,n+1} f(t_j, y_n(t_j)) \tag{1.54}$$

$$a_{j,n+1} = \begin{cases} n^{q+1} - (n - q)(n + 1)^q & \text{if } j = 0 \\ (n - j + 2)^{q+1} + (n - j)^{q+1} + 2(n - j + 1)^{q+1} & \text{if } 1 \leqslant j \leqslant n \\ 1 & \text{if } j = n + 1 \end{cases} \tag{1.55}$$

In (1.54) the preliminar approximation $y_h^p(t_{n+1})$ is named predictor and is given by (1.56), where $b_{j,n+1}$ is evaluated by (1.57).

$$y_h^p(t_{n+1}) = \sum_{k=0}^{m-1} \frac{t_{n+1}^k}{k!} y_0^{(k)} + \frac{1}{\Gamma(q)} \sum_{j=0}^{n} b_{j,n+1} f(t_j, y_n(t_j)) \tag{1.56}$$

$$b_{j,n+1} = \frac{h^q}{q}((n+1-j)^q - (n-j)^q) \tag{1.57}$$

As already shown in [69], both time domain numerical methods (Grünwald-Letnikov (1.45) and Adams-Bashforth-Moulton ((1.46)–(1.47)) have approximately the same exactness and a good approximation of the solution of a fractional-order dynamical system, as for the chaotic oscillators. However, as may be the case for integer-order chaotic oscillators, fractional-order ones require accomplishing the following theorems.

Definition Let us consider the general fractional-order system of n dimensions given by (1.58), in which the roots of evaluating $f(X) = 0$ become to be the equilibrium points, as described in Sect. 1.2 for the case of integer-order chaotic oscillators. In this case, $D^q(X) = (D^q x_1, D^q x_2, \ldots, D^q x_n)^T$, $X = (x_1, x_2, \ldots, x_n)^T \in R^n$.

$$D^q(X) = f(X) \tag{1.58}$$

Theorem 1.1 *A fractional-order system modeled by three state variables $n = 3$ is asymptotically stable at the equilibrium point equal to 0, if and only if $|arg(\lambda_i(J))| > q\pi/2$, $i = 1, 2, 3$. In this case J denotes the Jacobian matrix of $f(X)$, and λ_i are the eigenvalues of J [70].*

Theorem 1.2 *The equilibrium point O of a dynamical system described by (1.58) is unstable if and only if the order of q satisfies the condition imposed by (1.59), for at least one eigenvalue, where $\mathrm{Re}(\lambda)$ and $\mathrm{Im}((\lambda)$ denote the real and imaginary parts of λ [71].*

$$q > \frac{2}{\pi} \arctan \frac{|\mathrm{Im}((\lambda)|}{|\mathrm{Re}(\lambda)|} \tag{1.59}$$

Theorem 1.3 *For $n = 3$, if one of the eigenvalues $\lambda_1 < 0$ and the other two complex conjugated $|arg(\lambda_2)| = |arg(\lambda_2)| < q\pi/2$, then the equilibrium point O is called saddle point of index 2. If one of the eigenvalues $\lambda_1 > 0$ and the other two complex conjugated $|arg(\lambda_2)| = |arg(\lambda_3)| > q\pi/2$, then the equilibrium point O is called saddle point of index 1 [67].*

The algorithms that are used in this book to approach the solution of fractional-order chaotic oscillators are given in Algorithm 1 for the Grünwald-Letnikov method, and Algorithm 2 for the predictor-corrector Adams-Bashforth-Moulton method. These algorithms will be adapted in the next chapters to simulate different families of fractional-order chaotic oscillators and to develop their implementation using FPGAs. In Algorithm 1, one can see that from the Lorenz chaotic oscillator modeled by (1.1), its fractional-order version just changes the integer derivatives by fractional-order ones, and they are adapted to be solved applying the Grünwald-Letnikov method.

Algorithm 1 Grünwald-Letnikov method adapted to solve the fractional-order Lorenz chaotic oscillator

---

Require: f, q, y_0, T, T_0 y h.

1: **Output Variables**
2: y an array of $m \times N + 1$ real numbers that contain the approximate solutions
3: t an array of $N + 1$ real numbers that contain the solution from T_0 to T with an increase of h
4: **Internal Variables**
5: c_i the number of initial conditions
6: N the number of time steps that the algorithm is to consider
7: i, j variables used as indexes
8: c_1, c_2, c_3 arrays of $m \times N + 1$ real numbers that contain the values of the coefficients
9: $N = floor((T - T_0)/h$
10: $m = length(y_0)$
11: $y = zeros(m, N)$
12: **for** $(j = 1 to N)$ **do**
13:    $c_j[1, j] = \left(1 - \dfrac{1+q}{j}\right) c_{j-1}^q$
14: **end for**
15: **for** $(j = 1 \text{ to } N)$ **do**
16:    $x(i) = (\sigma(y(t_{i-1}) - x(t_{i-1})))h^{q_1} - \sum_{j=v}^{i} c_j^{(q_1)} x(t_{i-j}),$
       $y(i) = (x(i)(\rho - z(t_{k-1})) - y(t_{i-1}))h^{q_2} - \sum_{j=v}^{i} c_i^{(q_2)} y(t_{i-}),$
       $z(i) = (x(i)y(i) - \beta z(t_{i-1}))h^{q_3} - \sum_{j=v}^{i} c_i^{(q_3)} z(t_{i-j}),$
17: **end for**
18: **for** $(j = 1 \text{ to } N)$ **do**
19:    $y(j, 1) = x(j)$
20:    $y(j, 2) = y(j)$
21:    $y(j, 3) = z(j)$
22: **end for**

1.9 Simulation of the Fractional-Order Derivative $_0D_t^q y(t) = x(t)$

This section shows the evaluation of the fractional-order derivative $_0D_t^q y(t) = x(t)$ applying different fractional-order approximation methods.

Algorithm 2 Predictor-corrector Adams-Bashforth-Moulton method

---

Require: f, q, y_0, T, T_0 y h.

 1: **Output Variables**

 2: y an array of $m \times N + 1$ real numbers that contain the approximate solutions

 3: t an array of $N + 1$ real numbers that contain the solution from T_0 to T with an increase of h

 4: **Internal Variables**

 5: m the number of initial conditions

 6: N the number of time steps that the algorithm is to consider

 7: i, j variables used as indexes

 8: a, b arrays of $m \times N + 1$ real numbers that contain the values of the corrector and predictor, respectively

 9: p the predicted value

10: $N = floor((T - T_0)/h$

11: $m = length(y_0)$

12: $y = zeros(m, N)$

13: **for** $(j = 1 to N)$ **do**

14:     $b[1, j] = j^q - (j - 1)^\alpha$

15:     $a[1, j] = (j + 1)^{(q+1)} - 2j^{(q+1)} + (j - 1)^{(q+1)}$

16: **end for**

17: $y(:, 1) = y_0$

18: **for** $(i = 1 to N)$ **do**

19:     $p = y(:, i) + \dfrac{h^q}{\Gamma(q + 1)} \sum_i^{i=1} b[i] f(ih, y(i))$

20:     $y(:, i + 1) = y(:, i) + \dfrac{h^q}{\Gamma(q + 2)}(f(ih, p) + ((i - 1 - q)i^{j^q})f(0, y[0]) +$

     $\sum_{i=1}^{i} a[i] f(ih, y[i]))$

21: **end for**

1.9.1 Approximation Applying Grünwald-Letnikov Method

The fractional-order derivative can be approached applying the definition of Grünwald-Letnikov, where the solution is computed taking into account (1.60), in order to obtain the iterative formulae given in (1.61), which includes the short memory principle.

$$y(t_k) = f(y(t_k), t_k)h^q - \sum_{j=v}^{k} c_j^{(q)} y(t_{k-j}) \qquad (1.60)$$

$$x(t_k) = (x(t_{k-1})h^q - \sum_{j=v}^{k} c_j^{(q)} x(t_{k-j}) \qquad (1.61)$$

Equation (1.61) is evaluated straightforward, so that Fig. 1.13 shows the simulation results for the fractional order of the derivative in the range $0.1 \leq q \leq 1$, varied in steps of 0.1. In this case the solution does not use the short memory principle, it means that the exactness and convergency are maximum, but the approached

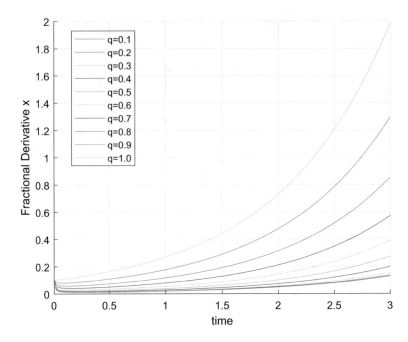

Fig. 1.13 Simulation of $_0D_t^q y(t) = x(t)$ applying Grünwald-Letnikov method varying the fractional order between $0.1 \leq q \leq 1$, and using: $h = 0.005$, initial condition of 0.1, time simulation between $0 \leq t \leq 3$ s, and performing 600 iterations

solution generates higher error than applying other methods, as shown below in the following subsection. Applying Grünwald-Letnikov method using the short memory principle, with lengths of 10, 20, 50, and 200, the approximations of the simulations are almost identical than using infinite memory for this simple fractional-order problem.

1.9.2 FDE12 Predictor-Corrector Method

The fractional-order derivative $_0D_t^q y(t) = x(t)$ can directly be simulated applying FDE12 available into MatLab^{TM} [72, 73]. In this case, using the same parameters from the Grünwald-Letnikov method, the solution is given in Fig. 1.14, which is more accurate. This can be taken in analogy to the error generated by one-step and multistep methods for solving integer-order chaotic oscillators, where Forward Euler is the lowest exact method. In fact, Table 1.4 highlights that the time-step must be much lower for Forward Euler to reach similar exactness than fourth-order Runge-Kutta or multistep methods. In the case of fractional-order chaotic oscillators, Grünwald-Letnikov method may require a lower time-step to reach similar accuracy than FDE12. This will be detailed in the following chapters.

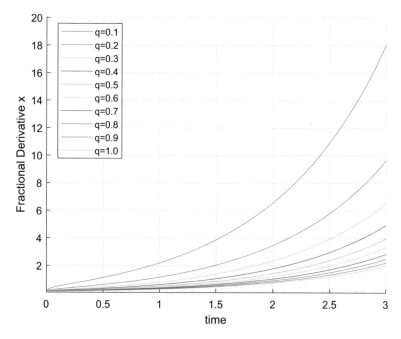

Fig. 1.14 Simulation of $_0D_t^q y(t) = x(t)$ applying FDE12 varying the fractional order between $0.1 \leq q \leq 1$, and using: $h = 0.005$, initial condition of 0.1, time simulation between $0 \leq t \leq 3\,\text{s}$, and performing 600 iterations

1.9.3   Adams-Bashforth-Moulton Method

The predictor-corrector approximation for solving integer-order chaotic oscillators is similar for fractional-order ones. In this case, the Adams-Bashforth-Moulton method detailed in [74] is applied herein to approach the solution of the fractional-order derivative [75]. In this case, the fractional-order differential equation given in (1.62) is considered.

$$D_t^q y(t) = f(y(t), t), \; y^{(k)}(0) = y_0^{(k)}, k = 0, 1, \ldots, m - 1 \qquad (1.62)$$

The discrete form of the method to obtain the corrector and iterative equation is expressed by (1.63), where $a_{j,n+1}$ is evaluated by (1.64), and using as predictor the discrete equation given in (1.65), in which $b_{j,n+1}$ is computed using (1.66). Figure 1.15 shows the simulation results for the fractional-orders varied in the range $0.1 \leq q \leq 1$.

$$y_h(t_{n+1}) = \sum_{k=}^{m-1} \frac{t_{n+1}^k}{k!} y_0^{(k)}$$

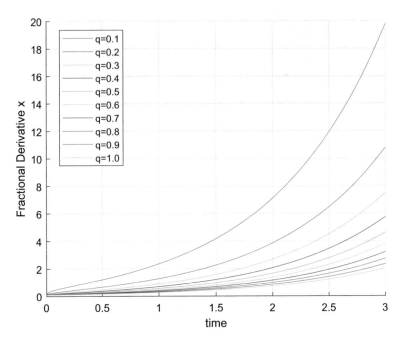

Fig. 1.15 Simulation of $_0D_t^q y(t) = x(t)$ applying Adams-Bashforth-Moulton method for fractional orders $0.1 \leq q \leq 1$, and using: $h = 0.005$, initial condition of 0.1, time simulation between $0 \leq t \leq 3\,\text{s}$, and performing 600 iterations

$$+ \frac{h^q}{\Gamma(\alpha + 2)} f(t_{n+1}, y_h^p(t_{n+1}))$$

$$+ \frac{h^q}{\Gamma(\alpha + 2)} \sum_{j=0}^{n} a_{j,n+1} f(t_j, y_n(t_j)) \tag{1.63}$$

$$a_{j,n+1} = \begin{cases} n^{q+1} - (n-q)(n+1)^q & \text{if } j = 0 \\ (n-j+2)^{q+1} + (n-j)^{q+1} + 2(n-j+1)^{q+1} & \text{if } 1 \leqslant j \leqslant n \\ 1 & \text{if } j = n+1 \end{cases} \tag{1.64}$$

$$y_h^p(t_{n+1}) = \sum_{k=0}^{m-1} \frac{t_{n+1}^k}{k!} y_0^{(k)} + \frac{1}{\Gamma(q)} \sum_{j=0}^{n} b_{j,n+1} f(t_j, y_n(t_j)) \tag{1.65}$$

$$b_{j,n+1} = \frac{h^q}{q}((n+1-j)^q - (n-j)^q) \tag{1.66}$$

As one can infer, the Adams-Bashforth-Moulton method approaches the solution of the fractional-order derivative in a similar exactness than it is done when applying the FDE12 method available into MatLab. The error generated applying the Grünwald-Letnikov method can be diminished varying the conditions of simulation, which in this section were the same as for the predictor-corrector methods. This means that for exactness issues, it will be much better to apply Adams-Bashforth-Moulton method than the Grünwald-Letnikov one, but as shown in Chap. 5, it will require a huge number of hardware resources, and if it requires huge lengths of memory resources, they may not be available in an FPGA for some special cases of fractional-order chaotic oscillators. Therefore, finding the appropriate parameters for both the Grünwald-Letnikov and Adams-Bashforth-Moulton methods, to generate low error, may be the best choice when trading exactness vs. hardware resources.

References

1. I. Petráš, *Fractional-Order Chaotic Systems* (Springer, Berlin, 2011), pp. 103–184
2. V.-T. Pham, S. Vaidyanathan, C. Volos, T. Kapitaniak, *Nonlinear Dynamical Systems with Self-excited and Hidden Attractors*, vol. 133 (Springer, Berlin, 2018)
3. H.K. Khalil, *Nonlinear Systems* (Prentice Hall, Englewood Cliffs, 1996)
4. P.A. Cook, *Nonlinear Dynamical Systems* (Prentice Hall, Englewood Cliffs, 1994)
5. H. Degn, A.V. Holden, L.F. Olsen, *Chaos in Biological Systems*, vol. 138 (Springer, New York, 2013)
6. V.H. Carbajal-Gomez, E. Tlelo-Cuautle, J.M. Muñoz-Pacheco, L.G. de la Fraga, C. Sanchez-Lopez, F.V. Fernandez-Fernandez, Optimization and CMOS design of chaotic oscillators robust to PVT variations. Integration **65**, 32–42 (2018)
7. A.D. Pano-Azucena, J. de Jesus Rangel-Magdaleno, E. Tlelo-Cuautle, A. de Jesus Quintas-Valles, Arduino-based chaotic secure communication system using multi-directional multi-scroll chaotic oscillators. Nonlinear Dynam. **87**(4), 2203–2217 (2017)
8. A.D. Pano-Azucena, E. Tlelo-Cuautle, J.M. Muñoz-Pacheco, L.G. de la Fraga, FPGA-based implementation of different families of fractional-order chaotic oscillators applying Grünwald–Letnikov method. Commun. Nonlinear Sci. Numer. Simul. **72**, 516–527 (2019)
9. A.A. Rezk, A.H. Madian, A.G. Radwan, A.M. Soliman, Reconfigurable chaotic pseudo random number generator based on FPGA. AEU-Int. J. Electron. Commun. **98**, 174–180 (2019)
10. O. Guillén-Fernández, A. Meléndez-Cano, E. Tlelo-Cuautle, J.C. Núñez-Pérez, J. de Jesus Rangel-Magdaleno, On the synchronization techniques of chaotic oscillators and their FPGA-based implementation for secure image transmission. PloS One **14**(2), e0209618 (2019)
11. C.K. Volos, D.A. Prousalis, S. Vaidyanathan, V.-T. Pham, J.M. Munoz-Pacheco, E. Tlelo-Cuautle, Kinematic control of a robot by using a non-autonomous chaotic system, in *Advances and Applications in Nonlinear Control Systems* (Springer, Berlin, 2016), pp. 1–17
12. T.S. Parker, L. Chua, *Practical Numerical Algorithms for Chaotic Systems* (Springer, New York, 2012)
13. E. Tlelo-Cuautle, L.G. de la Fraga, J. Rangel-Magdaleno, *Engineering Applications of FPGAs* (Springer, Berlin, 2016)
14. J.D. Lambert, *Computational Methods in Ordinary Differential Equations* (Wiley, Hoboken, 1973)
15. R.M. Corless, What good are numerical simulations of chaotic dynamical systems? Comput. Math. Appl. **28**(10–12), 107–121 (1994)
16. C. Varsakelis, P. Anagnostidis, On the susceptibility of numerical methods to computational chaos and superstability. Commun. Nonlinear Sci. Numer. Simul. **33**, 118–132 (2016)

17. E.N. Lorenz, Deterministic nonperiodic flow. J. Atmos. Sci. **20**(2), 130–141 (1963)
18. O.E. Rössler, An equation for continuous chaos. Phys. Lett. A **57**(5), 397–398 (1976)
19. G. Chen, T. Ueta, Yet another chaotic attractor. Int. J. Bifur. Chaos **9**(7), 1465–1466 (1999)
20. J. Lü, G. Chen, S. Zhang, Dynamical analysis of a new chaotic attractor. Int. J. Bifur. Chaos **12**(5), 1001–1015 (2002)
21. C. Liu, T. Liu, L. Liu, K. Liu, A new chaotic attractor. Chaos Solitons Fractals **22**(5), 1031–1038 (2004)
22. M.A. Zidan, A.G. Radwan, K.N. Salama, Controllable v-shape multiscroll butterfly attractor: system and circuit implementation. Int. J. Bifur. Chaos **22**(6), 1250143 (2012)
23. J.C. Sprott, Some simple chaotic flows. Phys. Rev. E **50**(2), R647 (1994)
24. M.W. Hirsch, S. Smale, R.L. Devaney, *Differential Equations, Dynamical Systems, and an Introduction to Chaos* (Academic, Cambridge, 2012)
25. A.D. Pano-Azucena, E. Tlelo-Cuautle, G. Rodriguez-Gomez, L.G. De la Fraga, FPGA-based implementation of chaotic oscillators by applying the numerical method based on trigonometric polynomials. AIP Adv. **8**(7), 075217 (2018)
26. D. Schleicher, Hausdorff dimension, its properties, and its surprises. Am. Math. Mon. **114**(6), 509–528 (2007)
27. A. Wolf, J.B. Swift, H.L. Swinney, J.A. Vastano, Determining Lyapunov exponents from a time series. Phys. D Nonlinear Phenomena **16**(3), 285–317 (1985)
28. V.H. Carbajal-Gómez, E. Tlelo-Cuautle, F.V. Fernández, L.G. de la Fraga, C. Sánchez-López, Maximizing Lyapunov exponents in a chaotic oscillator by applying differential evolution. Int. J. Nonlinear Sci. Numer. Simul. **15**(1), 11–17 (2014)
29. A. Silva-Juarez, G. Rodriguez-Gomez, L.G. de la Fraga, O. Guillen-Fernandez, E. Tlelo-Cuautle, Optimizing the Kaplan–Yorke dimension of chaotic oscillators applying de and PSO. Technologies **7**(2), 38 (2019)
30. G. Cardano, T.R. Witmer, *Ars Magna or the Rules of Algebra*. Dover Books on Advanced Mathematics (Dover, New York, 1968)
31. I. Petráš, *Fractional-Order Nonlinear Systems: Modeling, Analysis and Simulation* (Springer, New York, 2011)
32. A. Oustaloup, Fractional order sinusoidal oscillators: optimization and their use in highly linear FM modulation. IEEE Trans. Circuits Syst. **28**(10), 1007–1009 (1981)
33. A. Arenta, R. Caponetto, L. Fortuna, D. Porto, *Nonlinear Non-integer Order Circuits and Systems*. World Scientific Series on Nonlinear Science, Series A, vol. 38 (World Scientific, Singapore, 2002)
34. W.M. Ahmad, J.C. Sprott, Chaos in fractional-order autonomous nonlinear systems. Chaos, Solitons Fractals **16**(2), 339–351 (2003)
35. A.T. Azar, A.G. Radwan, S. Vaidyanathan, *Fractional Order Systems: Optimization, Control, Circuit Realizations and Applications* (Academic, Cambridge, 2018)
36. K. Rajagopal, S. Çiçek, A.J.M. Khalaf, V.-T. Pham, S. Jafari, A. Karthikeyan, P. Duraisamy, A novel class of chaotic flows with infinite equilibriums and their application in chaos-based communication design using DCSK. Z. Naturforsch. A **73**(7), 609–617 (2018)
37. C.K. Volos, S. Jafari, J. Kengne, J.M. Munoz-Pacheco, K. Rajagopal, Nonlinear *Dynamics and Entropy of Complex Systems with Hidden and Self-excited Attractors* (MDPI, Basel, 2019)
38. D. Baleanu, J.A.T. Machado, A.C.J. Luo, *Fractional Dynamics and Control* (Springer, New York, 2011)
39. C. Li, X. Liao, J. Yu, Synchronization of fractional order chaotic systems. Phys. Rev. E **68**(6), 067203 (2003)
40. R. Martínez-Guerra, C.A. Pérez-Pinacho, *Advances in Synchronization of Coupled Fractional Order Systems: Fundamentals and Methods* (Springer, Berlin, 2018)
41. A.T. Azar, S. Vaidyanathan, A. Ouannas, *Fractional Order Control and Synchronization of Chaotic Systems*, vol. 688 (Springer, Berlin, 2017)
42. A. Tepljakov, *Fractional-Order Modeling and Control of Dynamic Systems* (Springer, Berlin, 2017)

43. K. Rajagopal, S. Jafari, S. Kacar, A. Karthikeyan, A. Akgül, Fractional order simple chaotic oscillator with saturable reactors and its engineering applications. Inf. Technol. Control **48**(1), 115–128 (2019)
44. L.F. Ávalos-Ruiz, C.J. Zúñiga-Aguilar, J.F. Gómez-Aguilar, R.F. Escobar-Jiménez, H.M. Romero-Ugalde, FPGA implementation and control of chaotic systems involving the variable-order fractional operator with Mittag–Leffler law. Chaos Solitons Fractals **115**, 177–189 (2018)
45. K. Rajagopal, F. Nazarimehr, A. Karthikeyan, A. Srinivasan, S. Jafari, Fractional order synchronous reluctance motor: analysis, chaos control and FPGA implementation. Asian J. Control **20**(5), 1979–1993 (2018)
46. Z. Wei, A. Akgul, U.E. Kocamaz, I. Moroz, W. Zhang, Control, electronic circuit application and fractional-order analysis of hidden chaotic attractors in the self-exciting homopolar disc dynamo. Chaos Solitons Fractals **111**, 157–168 (2018)
47. E.-Z. Dong, Z. Wang, X. Yu, Z.-Q. Chen, Z.-H. Wang, Topological horseshoe analysis and field-programmable gate array implementation of a fractional-order four-wing chaotic attractor. Chin. Phys. B **27**(1), 010503 (2018)
48. K. Rajagopal, G. Laarem, A. Karthikeyan, A. Srinivasan, FPGA implementation of adaptive sliding mode control and genetically optimized PID control for fractional-order induction motor system with uncertain load. Adv. Differ. Equ. **2017**(1), 273 (2017)
49. K. Rajagopal, A. Karthikeyan, P. Duraisamy, Bifurcation analysis and chaos control of a fractional order portal frame with nonideal loading using adaptive sliding mode control. Shock. Vib. **2017**, Article ID 2321060, 14 (2017)
50. D.K. Shah, R.B. Chaurasiya, V.A. Vyawahare, K. Pichhode, M.D. Patil, FPGA implementation of fractional-order chaotic systems. AEU-Int. J. Electron. Commun. **78**, 245–257 (2017)
51. A. Karthikeyan, K. Rajagopal, Chaos control in fractional order smart grid with adaptive sliding mode control and genetically optimized PID control and its FPGA implementation. Complexity **2017**, Article ID 3815146, 18 (2017)
52. K. Oldham, J. Spanier, *The Fractional Calculus Theory and Applications of Differentiation and Integration to Arbitrary Order*, vol. 111 (Elsevier, Amsterdam, 1974)
53. S.S. Ray, *Fractional Calculus with Applications for Nuclear Reactor Dynamics* (CRC Press, Boca Raton, 2015)
54. O.M. Duarte, *Fractional Calculus for Scientists and Engineers* (Springer, Berlin, 2011), 114 pp.
55. F. Mainardi, *Fractional Calculus and Waves in Linear Viscoelasticity: An Introduction to Mathematical Models* (World Scientific, Singapore, 2010)
56. V.E. Tarasov, *Fractional Dynamics; Applications of the Fractional Calculus to Dynamics of Particles, Fields and Media* (Springer, Berlin, 2010), 522 pp.
57. D. Baleanu, Z.B. Günvec, M.J.A. Tenreiro, *New Trends in Nanotechnology and Fractional Calculus Applications* (Springer, Berlin, 2010), 544 pp.
58. C.-B. Fu, A.-H. Tian, Y.-C. Li, H.-T. Yau, Fractional order chaos synchronization for real-time intelligent diagnosis of islanding in solar power grid systems. Energies **11**(5), 1183 (2018)
59. Z. Gan, X. Chai, K. Yuan, Y. Lu, A novel image encryption algorithm based on LFT based S-boxes and chaos. Multimed. Tools Appl. **77**(7), 8759–8783 (2018)
60. V.P. Latha, F.A. Rihan, R. Rakkiyappan, G. Velmurugan, A fractional-order model for Ebola virus infection with delayed immune response on heterogeneous complex networks. J. Comput. Appl. Math. **339**, 134–146 (2018)
61. X. Lin, S. Zhou, H. Li, H. Tang, Y. Qi, Rhythm oscillation in fractional-order relaxation oscillator and its application in image enhancement. J. Comput. Appl. Math. **339**, 69–84 (2018)
62. K.S. Miller, B. Ross, *An Introduction to the Fractional Calculus and Fractional Differential Equations* (Wiley, Hoboken, 1993)
63. I. Podlubny, *Fractional Differential Equations: An Introduction to Fractional Derivatives, Fractional Differential Equations, to Methods of Their Solution and Some of Their Applications*. Mathematics in Science and Engineering (Elsevier, Amsterdam, 1999)

64. M. Caputo, Linear models of dissipation whose Q is almost frequency independent-II. Geophys. J. Int. **13**(5), 529–539 (1967)
65. L. Dorcak, J. Prokop, I. Kostial, Investigation of the properties of fractional-order dynamical systems, in *Proceedings of 11th International Conference on Process Control* (1994), pp. 19–20
66. I. Pan, S. Das, *Intelligent Fractional Order Systems and Control: An Introduction*, vol. 438 (Springer, Berlin, 2012)
67. W. Deng, J. Lü, Generating multi-directional multi-scroll chaotic attractors via a fractional differential hysteresis system. Phys. Lett. A **369**(5–6), 438–443 (2007)
68. N.J. Ford, A.C. Simpson, The numerical solution of fractional differential equations: speed versus accuracy. Numer. Algorithms **26**(4), 333–346 (2001)
69. Y. Chen, I. Petras, D. Xue, Fractional order control - a tutorial, in *2009 American Control Conference* (2009), pp. 1397–1411
70. I. Podlubny, *Fractional Differential Equations: An Introduction to Fractional Derivatives, Fractional Differential Equations, to Methods of Their Solution and some of Their applications*, vol. 198 (Elsevier, Amsterdam, 1998)
71. D. Cafagna, G. Grassi, On the simplest fractional-order memristor-based chaotic system. Nonlinear Dynam. **70**(2), 1185–1197 (2012)
72. R. Garrappa, Short tutorial: solving fractional differential equations by Matlab codes. Department of Mathematics, University of Bari (2014)
73. M.-F. Danca, N. Kuznetsov, Matlab code for Lyapunov exponents of fractional-order systems. Int. J. Bifurcation Chaos **28**(5), 1850067 (2018)
74. K. Diethelm, N.J. Ford, A.D. Freed, A predictor-corrector approach for the numerical solution of fractional differential equations. Nonlinear Dynam. **29**(1–4), 3–22 (2002)
75. J.M. Muñoz-Pacheco, E. Zambrano-Serrano, O. Félix-Beltrán, L.C. Gómez-Pavón, A. Luis-Ramos, Synchronization of PWL function-based 2d and 3d multi-scroll chaotic systems. Nonlinear Dynam. **70**(2), 1633–1643 (2012)

Chapter 2
FPAA-Based Implementation and Behavioral Descriptions of Autonomous Chaotic Oscillators

2.1 FPAA-Based Implementation of Autonomous Chaotic Oscillators

The first electronic implementations of autonomous chaotic oscillators were developed using operational amplifiers and diodes. The ODEs modeling the chaotic oscillators require integer-order integrators that can be implemented using different kinds of amplifiers and also they can be designed using complementary-metal-oxide-semiconductor (CMOS) technology of integrated circuits, as already summarized in [1]. More recently, some works show the optimization of chaotic oscillators and their robust design by performing an analysis on process, voltage, and temperature (PVT) variations, as shown in [2].

Fractional-order chaotic oscillators can be implemented using fractional-order integrators, which can be implemented by synthesizing fractances or fractors that emulate a fractional-order capacitor [3], as detailed in Chap. 4; or by approaching the fractional-order integrator by a Laplace transfer function applying frequency-domain methods. For the last case, the approached transfer function can be designed using traditional integer-order integrators, as shown in [4], where the authors introduced new alternatives for analog implementation of fractional-order integrators, differentiators, and proportional-integral-derivative (PID) controllers based on integer-order integrators. Having integer-order integrators, they can be implemented using operational amplifiers (opamps), and further, the opamp-based topologies can be transformed to be implemented with operational transconductance amplifiers (OTAs) and capacitors, as shown in [5]. That way, the resulting OTA-C topology can be designed with CMOS integrated circuit technology, as shown in [2]. This design process can be extended to implement fractional-order chaotic oscillators, which is detailed in the following chapters, where basically one should approach fractional-order integrators using fractances [3], or using integer-order integrators, as shown in [4]. Those topologies can be synthesized into an FPAA as described in [6], and as it is detailed herein.

© Springer Nature Switzerland AG 2020
E. Tlelo-Cuautle et al., *Analog/Digital Implementation of Fractional Order Chaotic Circuits and Applications*, https://doi.org/10.1007/978-3-030-31250-3_2

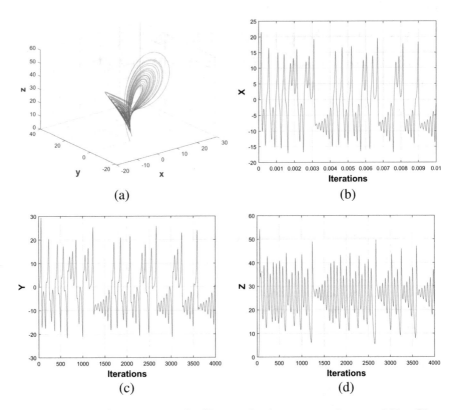

Fig. 2.1 (**a**) x, y, z phase-space portrait of Lorenz chaotic attractor, and state variables, (**b**) x, (**c**) y, and (**d**) z

Let us consider the Lorenz chaotic oscillator described by (1.1). By setting $\sigma = 10, \rho = 28$, and $\beta = 8/3$, the attractor and chaotic time series are shown in Fig. 2.1. The amplitude ranges of the state variables x, y, z are equal to $(-21.5; 21.5)$, $(-29.44; 29.44)$, and $(-54.26; 54.26)$, respectively. These ranges are higher than the voltage ranges supported by commercially available amplifiers and multipliers, so that the mathematical equations should be down-scaled to allow the implementation using electronic devices.

Using amplifiers and multipliers, the circuit realization of (1.1) is shown in Fig. 2.2, which consists of three opamps TL082 to perform the addition, subtraction, and integration operations at the same time, and two multipliers AD633 to implement the nonlinear functions xz and xy. The scaling operations are performed as follows: The circuit elements associated with σ, ρ, and β are scaled to 1 MΩ for the resistances in which $\sigma = 10$ is related to R_1 and R_2 for the state variables x and y, respectively. Defining $C_1 = C_2 = C_3 = 2\,\text{nF}$, the scaled $R_1 = 1\,\text{MΩ}/100\,\text{kΩ}$, and the time constant for the three integrators is established to be $1/R_1 C_1 = 5 \times 10^3$.

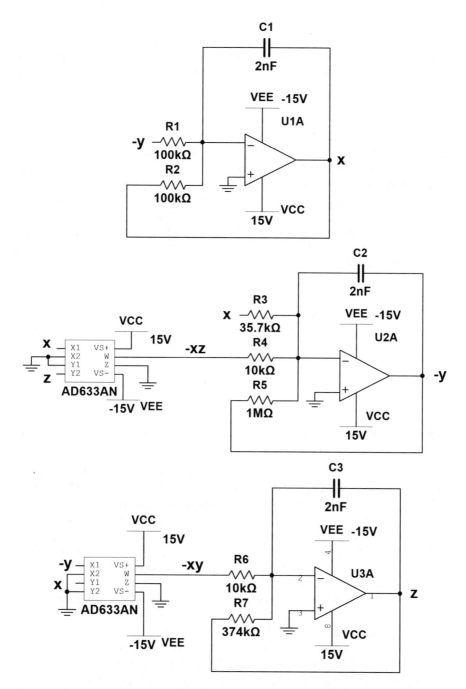

Fig. 2.2 Opamp-based implementation of Lorenz chaotic circuit showing the state variables: (**a**) x, (**b**) y, and (**c**) z

The second equation of Lorenz oscillator $\dot{y} = x(\rho - z) - y$, has $\rho = 28$, and is associated with R_3 and x, therefore the scaled value becomes $R_3 = 1\,\text{M}\Omega/35.7\,\text{k}\Omega$. The second input is associated with $-xz$ provided at the output of the multiplier AD633 and R_3 with unitary scaling, so that the equations are normalized to an amplitude of 0.1 V, this is the scale factor 100 of the multiplier, i.e., $R_4 = 1\,\text{M}\Omega/10\,\text{k}\Omega$. The third input has y with unitary value $R_5 = 1\,\text{M}\Omega/1\,\text{M}\Omega$. For the equation $\dot{z} = xy - \beta z$, it is associated with the output $-xy$ of the multiplier AD633 and R_6 with unitary scaling, i.e., $R_6 = 1\,\text{M}\Omega/10\,\text{k}\Omega$. The second input is associated with z and $\beta = 8/3$, leading to scale $R_7 = 1\,\text{M}\Omega/374\,\text{k}\Omega$.

The scaling discussed above accomplishes that all the state variables have a similar dynamic range so that their voltage ranges are within the supplies of the amplifiers and multipliers. The equivalent equations replacing the coefficients by R and C elements are now given by (2.1), which are obtained by analyzing Fig. 2.2. The simulation results are shown in Fig. 2.3, in which the passive circuit elements are set to: $R_1 = R_2 = 100\,\text{k}\Omega$, $R_3 = 35.7\,\text{k}\Omega$, $R_4 = R_6 = 10\,\text{k}\Omega$, $R_5 = 1\,\text{M}\Omega$, $R_7 = 374\,\text{k}\Omega$, and $C_1 = C_2 = C_3 = 2\,\text{nF}$.

$$
\begin{aligned}
\frac{dx}{dt} &= -\frac{1}{R_1 C_1}y + \frac{1}{R_2 C_1}x \\
\frac{dy}{dt} &= \frac{1}{R_3 C_2}x - \frac{1}{R_4 C_2}xz - \frac{1}{R_5 C_2}y \\
\frac{dz}{dt} &= -\frac{1}{R_6 C_3}xy + \frac{1}{R_7 C_3}z
\end{aligned}
\tag{2.1}
$$

2.1.1 Implementation Using AN231E04 FPAA

Field-programmable analog arrays (FPAAs) are analog signal processors, equivalent to the digital ones (FPGAs). FPAAs are electronic devices of specific purpose and having the characteristic of being reconfigurable electrically [6]. An FPAA can be used to implement a wide variety of analog functions, such as integration, derivation, pondered sum/subtraction, filtering, rectification, comparator, multiplication, division, analog-to-digital conversion, voltage references, signal conditioning, amplification, synthesis of nonlinear functions, and generation of arbitrary signals, among others [4].

The Anadigm QuadApex development board is an easy-to-use platform designed to allow fast prototyping to implement and test analog designs on the Anadigm-Apex FPAA silicon device. Such development board has the advantage of embedding a 32 bit PIC32 micro-controller and 4 FPAA devices. In this manner, it provides an extremely powerful platform that is biased with 3.3 V and includes an USB cable and is programmed using Anadigm Designer 2 EDA software, which generates C-code automatically. To implement chaotic oscillators, one must be aware that the dynamic voltage-range of the FPAA is ±3 V, therefore, scaling processes may be required according to desired results. The chaotic oscillators can be modeled

Fig. 2.3 Attractors of
Lorenz's chaotic circuit
(1 V/Div) in the: (**a**) $x - y$,
(**b**) $x - z$, and (**c**) $y - z$
portraits

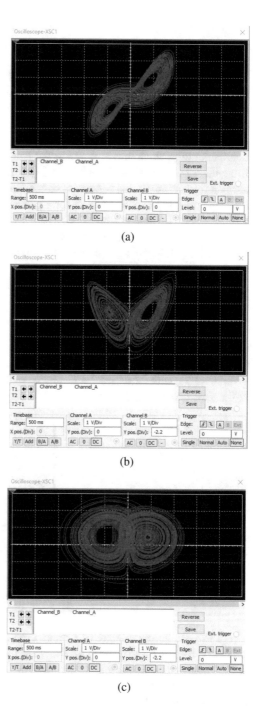

(a)

(b)

(c)

using the resources provided by the FPAA-EDA software tool, and afterwards, the developed model can be synthesized on the FPAA development board via USB interface [7].

The AN231E04 device is quite useful to accommodate nonlinear functions, linearization of the response of a sensor, and arbitrary waveform synthesis. The FPAA consists of a 2×2 matrix of fully Configurable Analog Blocks (CABs), which are surrounded by programmable interconnect resources and analog input/output cells with active elements. An on-chip clock generator block controls multiple nonoverlapping clock domains generated from an external and stable clock source. In some applications, an internal band-gap reference generator is used to create temperature compensated reference voltage levels. The inclusion of an 8×256 bit look-up table (LUT) enables waveform synthesis and handling of several nonlinear functions. As a whole system, the FPAA features seven configurable input/output structures, four have integrated differential amplifiers. There is also a single chopper stabilized amplifier that can be used by three of the seven output cells. The design process requires describing circuit functions that are represented as configurable analog modules (CAMs), which map onto portions of CABs. The EDA software and development board facilitates instant prototyping of any circuit that is implemented in the EDA tool [8].

As mentioned above, the ranges of the state variables of Lorenz chaotic oscillator shown in Fig. 2.1 cannot be synthesized into an FPAA because the allowed range of the FPAA AN231E04 is between ± 3 V. That way, one can choose the scaling factors given in (2.2), with $k_x = 5$, $k_y = 5$, and $k_z = 5$, thus updating the Lorenz equations to the ones given in (2.3).

$$\begin{cases} X = \dfrac{x}{k_x} \\ Y = \dfrac{y}{k_y} \\ Z = \dfrac{z}{k_z} \end{cases} \tag{2.2}$$

$$\begin{aligned} \dot{X} &= \sigma \left(\frac{k_y}{k_x} Y - X \right) \\ \dot{Y} &= \rho \frac{k_x}{k_y} X - Y - \frac{k_x k_z}{k_y} X Z \\ \dot{Z} &= \frac{k_x k_y}{k_z} X Y - \beta Z \end{aligned} \tag{2.3}$$

All the design parameters of Lorenz circuit are programmable but their values cannot be fixed with exactness, so that they include certain tolerances. In addition, one can introduce additional parameters to the mathematical model to diminish errors due to the programmable device. In this manner, the FPAA-based implementation is based on (2.4), where $G_1, G_2 \ldots G_5$ are experimentally adjusted. In addition to the multipliers, a CAM Sum/Difference Integrator is used to perform addition, subtraction, and integration operations. The CAM creates a summing integrator with up to three inputs, which may be either inverting or non-inverting to create sums and differences in the transfer function.

$$
\begin{aligned}
\dot{X} &= \alpha \left(\frac{k_y}{k_x} Y - X \right) \\
\dot{Y} &= G_1 \rho \frac{k_x}{k_y} X - G_2 Y - G_3 \frac{k_x k_z}{k_y} XZ \\
\dot{Z} &= G_4 \frac{k_x k_y}{k_z} XY - G_5 \beta Z
\end{aligned}
\tag{2.4}
$$

Each sampled input branch has a programmable integration constant. For instance, the transfer function given in (2.5) has three variables K_1, K_2, K_3 associated with the integration constants, and they are accompanied by three inputs branches $V_{Input1}, V_{Input2}, V_{Input3}$. The third term in (2.5) will only be implemented if the corresponding CAM Option Input is turned on. The sign of each term depends on the polarity selected for each input branch in the CAM Options. Other terms are added for non-inverting inputs and subtracted for inverting inputs. For example, the transfer function for this CAM configured with only two non-inverting inputs becomes $\frac{\Delta V_{out}}{\Delta t} = \pm K_1 V_{Input1} \pm K_2 V_{Input2}$, where ΔV_{out} is the change in the output voltage during one clock period, and Δt is the length of one clock period. The equivalent transfer function in the Laplace domain is given in (2.6), where the constants are replaced by capacitor values satisfying the following relations: $K_1 = \frac{f_c C_{inA}}{C_{int}}$, $K_2 = \frac{f_c C_{inB}}{C_{int}}$, and $K_3 = \frac{f_c C_{inC}}{C_{int}}$.

$$
\frac{\Delta V_{out}}{\Delta t} = \pm K_1 V_{Input1} \pm K_2 V_{Input2} \pm K_3 V_{Input3}
\tag{2.5}
$$

$$
V_{out}(s) = \frac{\pm K_1 V_{Input1}(s) \pm K_2 V_{Input2}(s) \pm K_3 V_{Input3}(s)}{s}
\tag{2.6}
$$

The integration constants are evaluated as $K = 1/RC$. One must take into account that when implementing these constants in Anadigm Designer 2, they must accomplish the available values in $1/\mu S$, not in $1/S$. Therefore, calculating K, it can be established to $K = 1 \times 10^{-6}/RC$ to get the CAM-parameters, as shown in Table 2.1. Figure 2.4 shows the seven configurable input/output structures IOCELL, where 5, 6, and 7 are configured as outputs and Bypass type, as shown also in Table 2.1.

The CAM is also used herein to perform the multiplication operations associated with XZ in the second and XY in the third equations of the Lorenz mathematical description given in (2.4). The CAM Multiplier listed in Table 2.1 has the transfer function given in (2.7), where M is related to a multiplication factor. This CAM of two inputs requires two frequencies to operate: Clock A and Clock B, having a relation requiring that Clock B is 16 times Clock A [8]. Table 2.2 gives the FPAA configuration for the Clock's to be used in the implementation of (2.4).

$$
V_{out} = M \cdot V_x \cdot V_y
\tag{2.7}
$$

Table 2.1 Configurable analog modules for the FPAA-based implementation of the Lorenz chaotic system

Block name	Options	Parameters	Clocks
SumIntegrator1	Output changes on: phase 1 Input1: non-inverting Input2: inverting Input3: inverting	Int. const. 1 (upper)[1/μs] 0.014 Int. const. 2 (middle)[1/μs] 0.05 Int. const. 3 (lower)[1/μs] 0.0005	ClockA 50 kHz (chip clock 1)
SumIntegrator2	Output changes on: phase 1 Input1: inverting Input2: inverting	Int. const. 1 (upper)[1/μs] 0.005 Int. const. 2 (lower)[1/μs] 0.005	ClockA 50 kHz (chip clock 1)
SumIntegrator3	Output changes on: phase 1 Input1: inverting Input2: inverting	Int. const. 1 (upper)[1/μs] 0.005 Int. const. 2 (lower)[1/μs] 0.00134	ClockA 50 kHz (chip clock 1)
Multiplier1	Output changes on: phase 1 Sample and hold: off	Multiplication factor 1.00	ClockA 50 kHz (chip clock 1) ClockB 800 kHz (chip clock 0)
Multiplier2	Output changes on: phase 1 Sample and hold: off	Multiplication factor 1.00	ClockA 50 kHz (chip clock 1) ClockB 800 kHz (chip clock 0)
GainHold1	Input sampling phase 1 Sample and hold: off	Gain 1.00	ClockA 50 kHz (chip clock 1)
GainHold2	Input sampling phase 1 Sample and hold: off	Gain 1.00	ClockA 50 kHz (chip clock 1)
IOCell5	I/O mode: output Output of variable x	Output type: bypass	
IOCell6	I/O mode: output Output of variable y	Output type: bypass	
IOCell7	I/O mode: output Output of variable z	Output type: bypass	

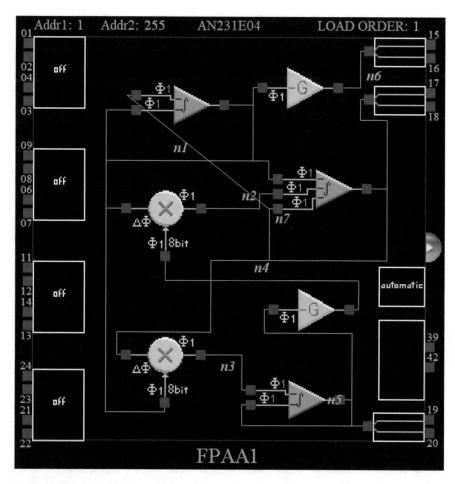

Fig. 2.4 FPAA-based implementation of Lorenz chaotic oscillator

Table 2.2 Frequencies of the Clocks of the FPAA

Master clock—ACLK (fc)	16 MHz
System clock 1 (sys1 = fc/20)	800 kHz
System clock 2 (sys2 = fc/1)	16 MHz
Clock 0 (sys1/1)	800 kHz
Clock 1 (sys1/16)	50 kHz

2.2 Simulating Chaotic Oscillators in Anadigm Designer Tool

Anadigm Designer 2 EDA tool allows us to quickly and easily construct complex analog circuits by selecting, placing, and wiring building block like sub-circuits that are referred to as CAMs, and which can be synthesized into an FPAA. Before the physical implementation, one can simulate the FPAA-based implementation of a

Fig. 2.5 Hardware configuration using Anadigm Designer software

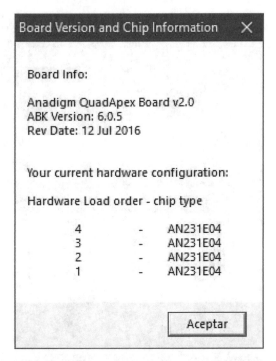

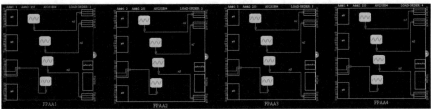

Fig. 2.6 Sketch of the four FPAAs embedded into AN231E04 Anadigm QuadApex

chaotic oscillator in Anadigm Designer software, where the simulation results can be viewed immediately using the oscilloscope option [9].

Figure 2.5 shows the information on the current hardware configuration displayed by Anadigm Designer software. Figure 2.6 shows the four FPAA AN231E04 embedded into Anadigm QuadApex development board highlighting a sinusoidal signal of 2 V_{pp} at the outputs of the FPAAs.

To simulate the scaled equations of Lorenz chaotic oscillator given in (2.4), the first step consists on estimating the required FPAA resources, in this case, one can use the resources of a single chip in AN231E04. Afterwards, one must perform three steps: select, adjust, and place the CAMs; connect the CAMs and IO cells with wires; and synthesize the configured data into the FPAA. The main goal in

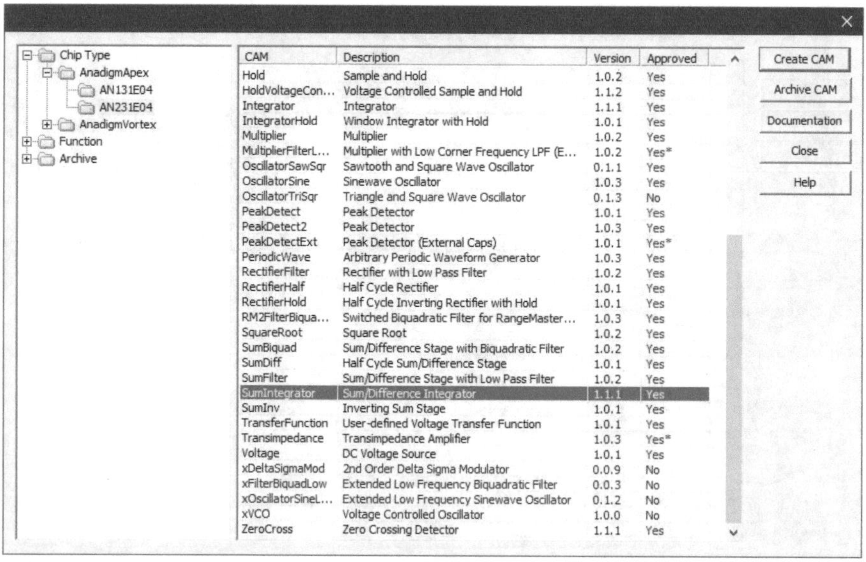
Chip Type
AnadigmApex
AN131E04
AN231E04
AnadigmVortex
Function
Archive
CAM
Description
Version
Approved
Hold
Sample and Hold
1.0.2
Yes
HoldVoltageCon...
Voltage Controlled Sample and Hold
1.1.2
Yes
Integrator
Integrator
1.1.1
Yes
IntegratorHold
Window Integrator with Hold
1.0.1
Yes
Multiplier
Multiplier
1.0.2
Yes
MultiplierFilterL...
Multiplier with Low Corner Frequency LPF (E...
1.0.2
Yes*
OscillatorSawSqr
Sawtooth and Square Wave Oscillator
0.1.1
Yes
OscillatorSine
Sinewave Oscillator
1.0.3
Yes
OscillatorTriSqr
Triangle and Square Wave Oscillator
0.1.3
No
PeakDetect
Peak Detector
1.0.1
Yes
PeakDetect2
Peak Detector
1.0.3
Yes
PeakDetectExt
Peak Detector (External Caps)
1.0.1
Yes*
PeriodicWave
Arbitrary Periodic Waveform Generator
1.0.3
Yes
RectifierFilter
Rectifier with Low Pass Filter
1.0.2
Yes
RectifierHalf
Half Cycle Rectifier
1.0.1
Yes
RectifierHold
Half Cycle Inverting Rectifier with Hold
1.0.1
Yes
RM2FilterBiqua...
Switched Biquadratic Filter for RangeMaster...
1.0.3
Yes
SquareRoot
Square Root
1.0.2
Yes
SumBiquad
Sum/Difference Stage with Biquadratic Filter
1.0.2
Yes
SumDiff
Half Cycle Sum/Difference Stage
1.0.1
Yes
SumFilter
Sum/Difference Stage with Low Pass Filter
1.0.2
Yes
SumIntegrator
Sum/Difference Integrator
1.1.1
Yes
SumInv
Inverting Sum Stage
1.0.1
Yes
TransferFunction
User-defined Voltage Transfer Function
1.0.1
Yes
Transimpedance
Transimpedance Amplifier
1.0.3
Yes*
Voltage
DC Voltage Source
1.0.1
Yes
xDeltaSigmaMod
2nd Order Delta Sigma Modulator
0.0.9
No
xFilterBiquadLow
Extended Low Frequency Biquadratic Filter
0.0.3
No
xOscillatorSineL...
Extended Low Frequency Sinewave Oscillator
0.1.2
No
xVCO
Voltage Controlled Oscillator
1.0.0
No
ZeroCross
Zero Crossing Detector
1.1.1
Yes
Create CAM
Archive CAM
Documentation
Close
Help

Fig. 2.7 Selection of CAMs using Anadigm designer

this section is the description of the FPAA-based implementation of Lorenz chaotic oscillator, as shown in Fig. 2.4, which corresponds to the synthesis of (2.4).

Using Anadigm Designer EDA tool, one starts with a blank design window, so that one can choose File $\rightarrow$ New and all contents of the main window will be cleared, and one is ready to select the CAMs, as listed in Fig. 2.7. From Lorenz equations, one must select the inverting or non-inverting inputs in the CAMs. The parameters are previously calculated for the three integrators and the frequencies of the Clocks A and B (see Table 2.1). Figure 2.8 shows the configuration of the CAM to implement each equation of Lorenz in (2.4).

The multipliers are configured as shown in Fig. 2.9, in which the CAM requires two clocks, with Clock B having 16 times Clock A. The required inverters with unity gain are configured in a similar way.

The Lorenz chaotic oscillator is an autonomous circuit but Anadigm Designer requires that all circuits have a signal generator. It can be added as a dummy block in order that the EDA tool can perform the simulation, as shown in Fig. 2.10. The CAMs implementing sample and hold operations mainly delay of one clock phase [8], thus avoiding zero delays. In Fig. 2.10 it can also be observed that there is an external connection between the output O07 and input IO3, in which the cables are interchanged to invert the signal. This can be performed to emulate an inverter when the resources of the FPAA are limited. The synthesis of the Lorenz chaotic oscillator in the FPAA will not require the CAMs implementing the delays. The simulation results provided by Anadigm Designer 2 are shown in Fig. 2.11.

The synthesis of (2.4) into the FPAA Anadigm QuadApex Development Board AN231E04 requires the configuration of the CAM shown in Table 2.1. Figure 2.4

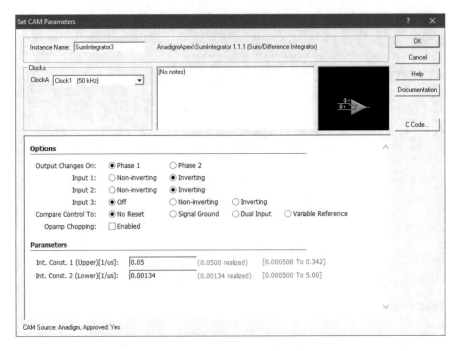

Fig. 2.8 CAM configuration to implement the three equations of Lorenz (2.4)

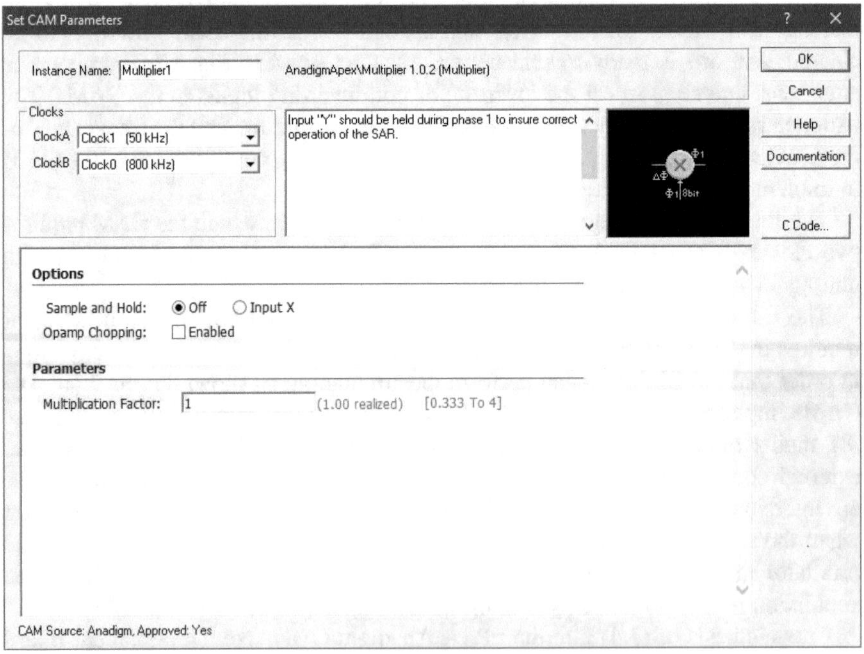

Fig. 2.9 CAM configuration to implement the multipliers for the operations xy and xz

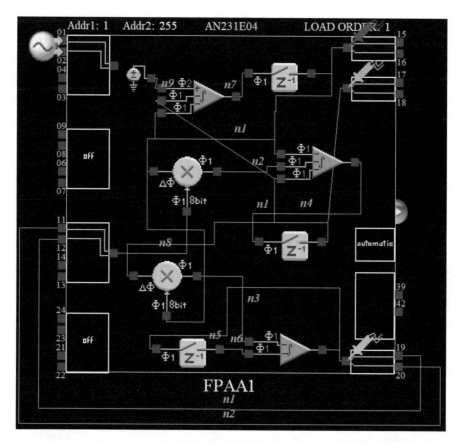

Fig. 2.10 Description of Lorenz chaotic oscillator in Anadigm Designer 2 including delay blocks (z^{-1}) to observe the state variables x, y, z

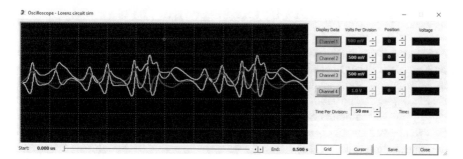

Fig. 2.11 Simulation of Lorenz chaotic oscillator in Andigm Designer 2 to observe the state variables x, y, z

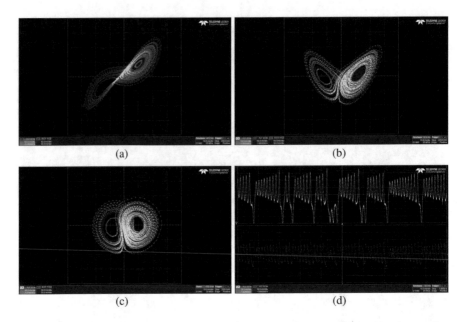

Fig. 2.12 Experimental observation of the chaotic attractors of the FPAA-based implementation of Lorenz circuit (50 mV/Div): (**a**) portrait $x - y$, (**b**) portrait $x - z$, (**c**) portrait $y - z$, and (**d**) time series x (200 mV/Div) with amplitude of 1.518 V, and y (50 mV/Div) with an amplitude of 1.666 V

shows the final design that does not include the delay blocks shown in Fig. 2.10. The FPAA-based implementation is measured using an oscilloscope to observe the experimental attractors and chaotic time series, which are shown in Fig. 2.12.

2.3   Implementing Chaotic Oscillator Using FPGAs

Nowadays, there are various engineering applications based on field-programmable gate arrays (FPGAs) because they exploit their fast prototyping and reconfigurability advantages. Some analog circuits and any digital circuit can be implemented in an FPGA, so the possibilities of applications are endless. This chapter details the description of mathematical models to VHDL programming. The VHDL code can be synthesized on FPGAs to develop applications as data encryption and secure communication systems [10].

Before implementing the chaotic system into an FPGA, the mathematical model like the Lorenz system (1.1) must be discretized according to a selected numerical method, as already shown in [11]. For example, applying Forward Euler, as shown in Chap. 1, the discretized equations are given in (2.8), where one can identify digital blocks associated with multipliers, adders, and subtractors. The iterations can be controlled by multiplexers to begin with initial conditions and a finite state machine to control the iterations. Chapter 1 shows other one-step and multistep methods to simulate chaotic oscillators.

$$x_{n+1} = x_n + h[\sigma(y_n - x_n)]$$
$$y_{n+1} = y_n + h[x_n(\rho - z_n) - yn]$$ \hspace{1cm} (2.8)
$$z_{n+1} = z_n + h[x_n y_n - \beta z_n]$$

In this chapter the digital blocks are described under the VHDL language, which is based on the numeric representation to perform computer arithmetic among the blocks. Fixed-point notation is very easy to synthesize into an FPGA, in which one bit is used to represent the positive or negative sign and one can establish the number of bits to represent the integer and fractional numbers, according to the ranges of the state variables that are known from its numerical simulation. According to the simulation results for the Lorenz oscillator given in Chap. 1, in this chapter a fixed-point representation using 32 bit is used. One bit is used to represent the sign, 11 bit for the integer part, and 20 bit for the fractional part. The blocks identified in (2.8) are sketched in Fig. 2.13, and this chapter details their VHDL description. In those blocks a clock (clk) and reset (rst) pins are used to avoid the use of shift registers after each operation and then save digital resources. Other specific blocks like single constant multiplier (SCM), register, and counter are sketched in Fig. 2.14. In particular, SCM blocks are required to multiply a state variable with a constant

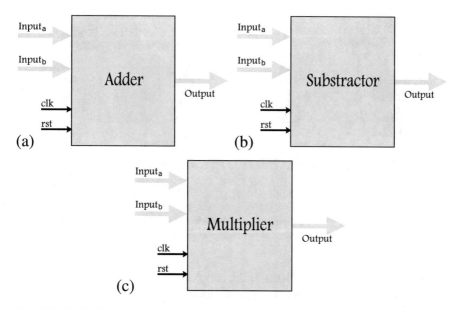

Fig. 2.13 Digital blocks with two input-buses describing: (**a**) an adder, (**b**) subtractor, and (**c**) multiplier. All including clk and rst pins

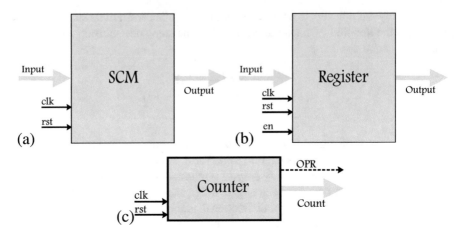

Fig. 2.14 Special blocks: (**a**) single constant multiplier (SCM to multiply a variable by h, σ, ρ, and β), (**b**) parallel-parallel shift register, and (**c**) counter to control the iterations, requiring 7 clock cycles to evaluate each iteration of (2.8)

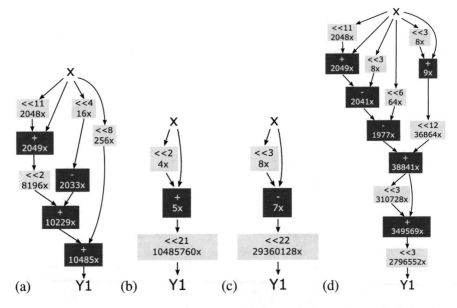

Fig. 2.15 Sketching the SCMs to multiply a variable by: (**a**) h, (**b**) σ, (**c**) ρ, and (**d**) β; using adders, subtraction, and shift operations to reduce hardware resources

value such as $h = 0.01$, $\sigma = 10$, $\rho = 28$, and $\beta = 8/3$, and then reduce hardware resources. Figure 2.15 shows the details of the SCM for each constant, as already done in [12].

Figure 2.14b shows a parallel-parallel shift register that is used to hold the result at the actual iteration in evaluating (2.8), and which are counted by the block counter

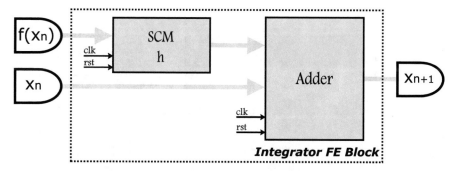

Fig. 2.16 General description of the hardware implementation of Forward Euler using an SCM to simulate a chaotic oscillator

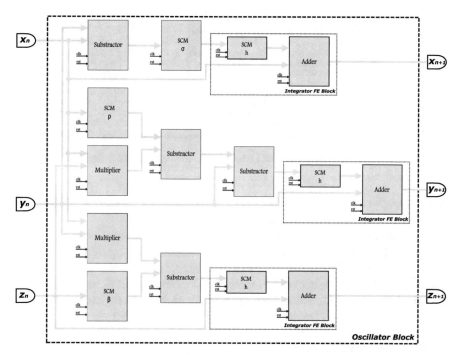

Fig. 2.17 Details of applying Forward Euler method to simulate Lorenz chaotic oscillator, showing the three state variables and other SCMs to multiply a state variable by the constants σ, ρ, and β

shown in Fig. 2.14c. In the Lorenz chaotic oscillator the final result at each iteration requires 7 clock cycles in the FPGA. Figure 2.16 shows the general description of Forward Euler, which is based on the iterative formulae $x_{n+1} = x_n + hf(x_n)$. As one sees, an SCM can be used to multiply the constant h by $f(x_n)$.

The details of the hardware implementation of Lorenz chaotic oscillator are shown in Fig. 2.17, where one can appreciate the state variables at the previous iteration (x_n, y_n, z_n), which are processed by adders, subtractors, multipliers, and

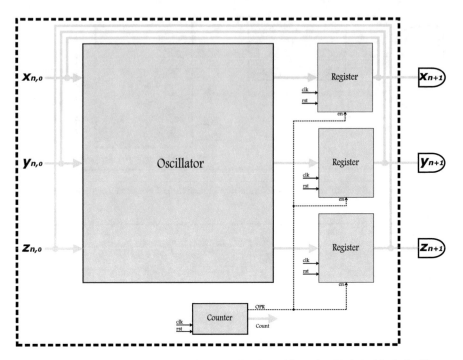

Fig. 2.18 Complete block description of Lorenz chaotic oscillator, in which the block Oscillator is described in Fig. 2.17

SCM blocks to reduce hardware resources. The outputs are the state variable values at the next iteration $(x_{n+1}, y_{n+1}, z_{n+1})$.

The complete Lorenz chaotic oscillator ready to be synthesized on the FPGA is shown in Fig. 2.18, where it can be seen that a counter, three registers, and a block named Oscillator are included. The block Oscillator is detailed in Fig. 2.17, and the Counter controls the iterations to process the data from the previous iteration n to the next iteration $n + 1$, and it takes seven clock cycles. The registers are used to set the initial conditions (x_0, y_0, z_0), so that at $n = 0$, the first iteration provides the values of (x_1, y_1, z_1).

The experimental attractors measured from the FPGA-based implementation of Lorenz chaotic oscillator using Forward Euler are shown in Fig. 2.19. The oscilloscope Teledyne Lecroy 2.5 GHz was used to observe 16 bit using a digital-to-analog converter, so that the 32 bit used in the FPGA implementation are truncated as detailed in the next sections, to select 16 bits that better show the experimental attractor. If the chaotic oscillator is implemented in an FPGA but now using the fourth-order Runge-Kutta method, the resulting experimental attractors are observed as shown in Fig. 2.20. As it can be observed, both attractors are quite similar in their phase-space portraits; however, one must recall that Forward Euler method requires a smaller step-size to provide similar accuracy than the fourth-order Runge-Kutta method, as shown in Chap. 1.

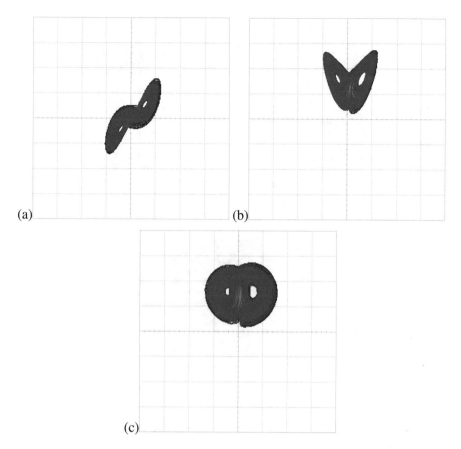

Fig. 2.19 Phase-space portraits of the experimental attractors of the FPGA-based Lorenz chaotic oscillator: (**a**) X–Y , (**b**) X–Z, and (**c**) Y–Z views with axes 2 V–2 V/division in the oscilloscope

2.4   VHDL Descriptions for Implementing Chaotic Oscillators

FPGAs are programmable semiconductor devices that are based on a matrix of configurable logic blocks (CLBs) connected through programmable interconnects. FPGAs can be reconfigured to modify or improve the design according to the application requirements. Each FPGA vendor designs a reconfigurable architecture, for example based on CLBs, logic cells, or logic elements, and including special hardware blocks like random access memory (RAM) or digital signal processor (DSP), which are commonly used devices, while their incorporation improves the resource utilization and maximum frequency of operation. For instance, static memory is the most widely used method for configuring FPGAs. In [13] one can find a brief introduction to Vivado from Xilinx, Quartus II from Altera, and Active-HDL.

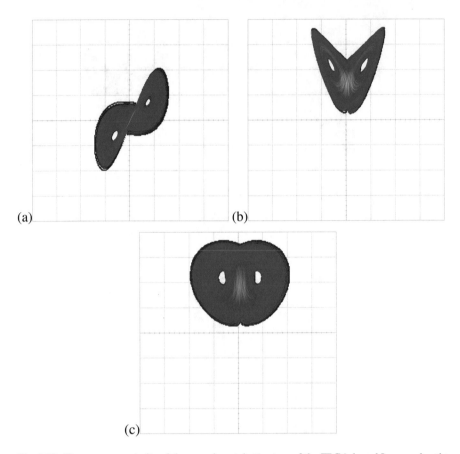

Fig. 2.20 Phase-space portraits of the experimental attractors of the FPGA-based Lorenz chaotic oscillator using the fourth-order Runge–Kutta method: (**a**) X–Y, (**b**) X–Z, and (**c**) Y–Z views with axes 2 V–2 V/division in the oscilloscope

In 1986 the Institute of Electrical and Electronics Engineers (IEEE) standardized the first hardware description language, VHDL, through the 1076 and 1164 standards. VHDL is technology/vendor independent, and VHDL codes are portable and reusable. VHDL is a structured language in which each description of a digital block should have three main blocks: Libraries, Entity, and Architecture. The main standard libraries for logic and arithmetic descriptions are for example:

```
library IEEE;
use IEEE. std_logic_1164 . all ;
use IEEE. std_logic_unsigned . all ;
use IEEE. std_logic_arith . all ;
use IEEE. numeric_std . all ;
```

These libraries and others like Unsigned and arith, which were developed by Synopsys Inc. under c language, will be used in the following descriptions of the blocks that will be used for the implementation of chaotic oscillators. The Entity can be seen as a black box that describes the inputs and outputs as ports, which can be unidirectional or bidirectional, and described as signals of type in, type out, type in/out, and also the option of type buffer that allows internal feedbacks inside the entity. The data type that is used herein is of bit type. Other types are: Boolean (take the values true or false), Integer (cover all integer values), and std_logic (allows nine values: U Unitialized, X Unknown, 0 Low, 1 High, Z High impedance, W Weak unknown, L Weak low, H Weak high, and '-' Don't care).

Listing 2.1 shows the entity description of a two inputs adder (a,b), one output (out1), including the pins for clock (clk) and reset (rst), and its architecture providing the sum of two signed operands. Listing 2.2 shows the entity and architecture descriptions of a two inputs subtractor, also including clk and rst pins. Listing 2.3 shows the entity and architecture descriptions of a two inputs multiplier, including clk and rst pins. One can see the use of an auxiliary signal in the architecture description, it allows 64 bit to the output, which is truncated to 32 bit in order that the output $out1$ uses the bits 51 down to 20, and the remaining are deleted. This process is done observing that those 32 bit are the ones that represent the dominant dynamical behavior of the chaotic data.

```
 1  ------------------------------------------------------------
 2  ---Project: Adder of two inputs of n bits              ---
 3  ---Instructions: Assign the inputs correctly.          ---
 4  ---     Inputs: clk - according to FPGA Model, example 50 MHz ---
 5  ---            rst - put a button/switch of FPGA       ---
 6  ---            a,b - logic vectors of n bits with sign ---
 7  ---     Output: out1- Gives the arithmetic sum of a+b in a ---
 8  ---                   logic vector                     ---
 9  ------------------------------------------------------------
10  library ieee;
11  use ieee.std_logic_1164.all;
12  use ieee.numeric_std.all;
13
14  Entity adder2 is
15     generic(n: integer:= 32);
16     port (
17     clk,rst :in std_logic;
18     a,b     :in std_logic_vector(n-1 downto 0);
19     out1    :out std_logic_vector(n-1 downto 0));
20  end adder2;
21
22  Architecture arch of adder2 is
23  begin
24     secuencial: process(rst,clk)
25     begin
26       if rst = '0' then
27         out1 <= (others => '0');
28       elsif (rising_edge(clk)) then
29         out1 <= std_logic_vector(signed(a)+signed(b));
30       end if;
31     end process;
32  end arch;
```

Listing 2.1 VHDL description of the ADDER block using 32 bit

```
1    -------------------------------------------------------------
2    ---Project: Subtractor of two inputs of n bits            ---
3    ---Instructions: Assign the inputs correctly.             ---
4    ---    Inputs: clk - according to FPGA Model, example 50 MHz ---
5    ---            rst - put a button/switch of FPGA          ---
6    ---            a,b - logic vectors of n bits with sign    ---
7    ---    Output: out1- Gives the arithmetic subtraction of a-b ---
8    ---                  in a logic vector                    ---
9    -------------------------------------------------------------
10   library ieee;
11   use ieee.std_logic_1164.all;
12   use ieee.numeric_std.all;
13
14   entity substractor2 is
15     generic(n: integer:= 32);
16     port (
17     clk,rst :in std_logic;
18     a,b     :in std_logic_vector(n-1 downto 0);
19     out1    :out std_logic_vector(n-1 downto 0));
20   end substractor2;
21
22   architecture arch of substractor2 is
23   begin
24     secuential: process(rst,clk)
25     begin
26       if rst = '0' then
27         out1 <= (others => '0');
28       elsif (rising_edge(clk)) then
29         out1 <= std_logic_vector(signed(a)-signed(b));
30       end if;
31     end process;
32   end arch;
```

Listing 2.2 VHDL description of the SUBTRACTOR using 32 bit

```
1    -------------------------------------------------------------
2    ---Project: Multiplier of two inputs of n bits            ---
3    ---Instructions: Assign the inputs correctly.             ---
4    ---    Inputs: clk - according to FPGA Model, example 50 MHz ---
5    ---            rst - put a button/switch of FPGA          ---
6    ---            a,b - logic vectors of n bits with sign    ---
7    ---    Output: out1- Gives the arithmetic multiplication of ---
8    ---                  a*b in a logic vector                ---
9    -------------------------------------------------------------
10   library ieee;
11   use ieee.std_logic_1164.all;
12   use ieee.numeric_std.all;
13
14   entity multiplier2 is
15     generic(n: integer:= 32);
16     port (
17     clk,rst :in std_logic;
18     a,b     :in std_logic_vector(n-1 downto 0);
19     out1    :out std_logic_vector(n-1 downto 0));
20   end multiplier2;
21
22   architecture arch of multiplier2 is
23   signal aux: signed(n+n-1 downto 0):=(others => '0');
24   begin
25     secuential: process(rst,clk)
26     begin
27       if rst ='0' then
28         out1<=(others =>'0');
29       elsif (rising_edge(clk)) then
30         aux <= signed(a)*signed(b);
```

```
31          out1 <= std_logic_vector(aux(51 downto 20));
32       end if;
33    end process;
34 end arch;
```

Listing 2.3   VHDL description of the MULTIPLIER using 32 bit

Apart from the standard blocks described above, like the adder, subtractor, and multiplier, other special blocks are required to reduce hardware resources when synthesizing the VHDL code into the FPGA. The single constant multiplier (SCM) block reduces the hardware of a standard multiplier, and it uses adders, subtractors, and shift registers. Listing 2.4 shows the VHDL description of the SCM used to multiply a state variable with the time-step $h = 0.01$ of the numerical method. In this case, this SCM block has one input $in1$ and one output $out1$, including clk and rst pins. According to Fig. 2.15, one requires to use special signals that are declared in the architecture. They are used to perform shifts and sums and the symbol & is used to concatenate the bits, and the output is truncated to 32 bit as described in the following sections.

Listings 2.5–2.7 show the VHDL descriptions of the other SCM required to multiply a state variable by the constants σ, ρ, and β, respectively. They require auxiliary signals in their architecture descriptions to accomplish the block operations shown in Fig. 2.15.

```
 1 -----------------------------------------------------------------
 2 ---Project: Single Constant Multiplier h = 0.01            ---
 3 ---Instructions: Assign the inputs correctly.             ---
 4 ---    Inputs: clk - according to FPGA Model, example 50 MHz ---
 5 ---            rst - put a button/switch of FPGA          ---
 6 ---            in1 - logic vector of n bits with sign     ---
 7 ---    Output: out1- Gives the arithmetic multiplication of ---
 8 ---                  in1*h in a logic vector               ---
 9 -----------------------------------------------------------------
10 library ieee;
11 use ieee.std_logic_1164.all;
12 use ieee.numeric_std.all;
13
14 entity scm_h is
15     port (
16     clk,rst : in std_logic;
17     in1     : in std_logic_vector(31 downto 0);
18     out1    : out std_logic_vector(31 downto 0));
19 end scm_h;
20
21 architecture arch of scm_h is
22 -- 1. Creation of signals to use in the mapping
23 signal w1, w2048, w2049, w16, w2033 : signed (51 downto 0);
24 signal w256, w2289, w8196, w10485   : signed (51 downto 0);
25 begin
26 -- 2. Adjust the size of the input in1 to 56 bits
27 --     by applying the resize instruction
28     w1 <= resize(signed(in1), w1'length);
29 -- 3. The shifts and sums of the signals are made
30 --     to obtain the correct result.
31 --     The symbol "&" is used to concatenate
32     w2048    <= w1(40 downto 0)&"00000000000";
33     w2049    <= w1 + w2048;
34     w16      <= w1(47 downto 0)&"0000";
35     w2033    <= w2049 - w16;
36     w256     <= w1(43 downto 0)&"00000000";
```

```
37      w2289    <= w2033 + w256;
38      w8196    <= w2049(49 downto 0)&"00";
39      w10485   <= w8196 + w2289;
40  -- 4. Sequential part
41    Process(clk, rst)
42    begin
43      if rst ='0' then
44        out1 <= (others=>'0');
45      elsif (clk'event and clk = '1') then
46  -- Select the output bits according to format used (i.e. 12.20)
47        out1 <= std_logic_vector(w10485(51 downto 20));
48      end if;
49    end process;
50  end arch;
```

Listing 2.4 VHDL description of the single constant multiplier (SCM) to multiply *h* using 32 bit

```
1   ----------------------------------------------------------------
2   ---Project: Single Constant Multiplier sigma s = 10         ---
3   ---Instructions: Assign the inputs correctly.              ---
4   ---    Inputs: clk - according to FPGA Model, example 50 MHz ---
5   ---            rst - put a button/switch of FPGA            ---
6   ---            in1 - logic vector of n bits with sign       ---
7   ---    Output: out1- Gives the arithmetic multiplication of ---
8   ---                  in1*s in a logic vector                ---
9   ----------------------------------------------------------------
10  library ieee;
11  use ieee.std_logic_1164.all;
12  use ieee.numeric_std.all;
13
14  entity scm_10 is
15    port (
16    clk,rst : in std_logic;
17    in1   : in std_logic_vector(31 downto 0);
18    out1  : out std_logic_vector(31 downto 0));
19  end scm_10;
20
21  architecture arch of scm_10 is
22  -- 1. Creation of signals to use in the mapping
23  signal w1, w4, w5, w10485760: signed (51 downto 0);
24  begin
25  -- 2. Adjust the size of the input in1 to 56 bits
26  --     by applying the resize instruction
27    w1 <= resize(signed(in1), w1'length);
28  -- 3. The shifts and sums of the signals are made
29  --     to obtain the correct result.
30  --     The symbol "&" is used to concatenate
31      w4    <= w1(49 downto 0)&"00";
32      w5    <= w1 + w4;
33  w10485760<= w5(30 downto 0)&"000000000000000000000";
34  -- 4. Sequential part
35    Process(clk, rst)
36    begin
37      if rst ='0' then
38        out1<=(others=>'0');
39      elsif (clk'event and clk='1') then
40  -- Select the output bits according to format used (i.e. 12.20)
41        out1<=std_logic_vector(w10485760(51 downto 20));
42      end if;
43    end process;
44  end arch;
```

Listing 2.5 VHDL description of the SCM to multiply σ using 32 bit

```vhdl
1    ------------------------------------------------------------------
2    ---Project: Single Constant Multiplier rho r = 28              ---
3    ---Instructions: Assign the inputs correctly.                  ---
4    ---    Inputs: clk - according to FPGA Model, example 50 MHz ---
5    ---            rst - put a button/switch of FPGA              ---
6    ---            in1 - logic vector of n bits with sign          ---
7    ---    Output: out1- Gives the arithmetic multiplication of    ---
8    ---                   in1*r in a logic vector                  ---
9    ------------------------------------------------------------------
10   library ieee;
11   use ieee.std_logic_1164.all;
12   use ieee.numeric_std.all;
13
14   entity scm_28 is
15     port (
16     clk,rst : in std_logic;
17     in1   : in std_logic_vector(31 downto 0);
18     out1  : out std_logic_vector(31 downto 0));
19   end scm_28;
20
21   architecture arch of scm_28 is
22   -- 1. Creation of signals to use in the mapping
23   signal w1, w8, w7, w29360128: signed (51 downto 0);
24   begin
25   -- 2. Adjust the size of the input in1 to 56 bits
26   --    by applying the resize instruction
27       w1 <= resize(signed(in1), w1'length);
28   -- 3. The shifts and sums of the signals are made
29   --    to obtain the correct result.
30   --    The symbol "&" is used to concatenate
31       w8      <= w1(48 downto 0)&"000";
32       w7      <= w8 - w1;
33     w29360128<= w7(29 downto 0)&"000000000000000000000000";
34   -- 4. Sequential part
35     Process(clk, rst)
36     begin
37       if rst ='0' then
38         out1<=(others=>'0');
39       elsif (clk'event and clk='1') then
40   -- Select the output bits according to format used (i.e. 12.20)
41         out1<=std_logic_vector(w29360128(51 downto 20));
42       end if;
43     end process;
44   end arch;
```

Listing 2.6   VHDL description of the SCM to multiply ρ using 32 bit

```vhdl
1    ------------------------------------------------------------------
2    ---Project: Single Constant Multiplier beta b = 8/3            ---
3    ---Instructions: Assign the inputs correctly.                 ---
4    ---    Inputs: clk - according to FPGA Model, example 50 MHz ---
5    ---            rst - put a button/switch of FPGA              ---
6    ---            in1 - logic vector of n bits with sign          ---
7    ---    Output: out1- Gives the arithmetic multiplication of    ---
8    ---                   in1*b in a logic vector                  ---
9    ------------------------------------------------------------------
10   library ieee;
11   use ieee.std_logic_1164.all;
12   use ieee.numeric_std.all;
13
14   entity scm_8_3 is
15     port (
16     clk,rst : in std_logic;
17     in1    : in std_logic_vector(31 downto 0);
18     out1   : out std_logic_vector(31 downto 0));
```

```
19   end scm_8_3;
20
21   architecture arch of scm_8_3 is
22   -- 1. Creation of signals to use in the mapping
23   signal w1, w2048, w2049, w8, w2041, w64, w1977, w9   :
24                   signed (51 downto 0);
25   signal w36864, w38841, w310728, w349569, w2796552   :
26                   signed (51 downto 0);
27   begin
28   -- 2. Adjust the size of the input in1 to 56 bits
29   --    by applying the resize instruction
30      w1 <= resize(signed(in1), w1'length);
31   -- 3. The shifts and sums of the signals are made
32   --    to obtain the correct result.
33   --    The symbol "&" is used to concatenate
34      w2048    <= w1(40 downto 0)&"00000000000";
35      w2049    <= w2048 + w1;
36      w8       <= w1(48 downto 0)&"000";
37      w2041    <= w2049 - w8;
38      w64      <= w1(45 downto 0)&"000000";
39      w1977    <= w2041 - w64;
40      w9       <= w8 + w1;
41      w36864   <= w9(39 downto 0)&"000000000000";
42      w38841   <= w1977 + w36864;
43      w310728  <= w38841(48 downto 0)&"000";
44      w349569  <= w310728 + w38841;
45      w2796552 <= w349569(48 downto 0)&"000";
46   -- 4. Sequential part
47     Process(clk, rst)
48     begin
49       if rst ='0' then
50         out1<=(others=>'0');
51       elsif (clk'event and clk='1') then
52   -- Select the output bits according to format used (i.e. 12.20)
53         out1<=std_logic_vector(w2796552(51 downto 20));
54       end if;
55     end process;
56   end arch;
```

Listing 2.7 VHDL description of the SCM to multiply β using 32 bit

Another block is the register that can have the length of n bit, but herein it is described using 32 bit. It includes clk and rst pins, a pin *en* to enable the register and the access to read the d logic vectors provided at the output.

```
1    ------------------------------------------------------------
2    ---Project: Shift Register of n bits                    ---
3    ---Instructions: Assign the inputs correctly.          ---
4    ---    Inputs: clk - according to FPGA Model, example 50 MHz ---
5    ---            rst - put a button/switch of FPGA       ---
6    ---            en - is the input that enable the register ---
7    ---            d - logic vectors of n bits             ---
8    ---    Output:   q - Gives d when en is enabled        ---
9    ------------------------------------------------------------
10   library ieee;
11   use ieee.std_logic_1164.all;
12
13   entity register1 is
14     generic(n      : integer := 32;
15          initial : std_logic_vector := x"0001999A");
16     port (
17     clk,rst,en  :in std_logic;
18     d           :in std_logic_vector(n-1 downto 0);
19     q           :out std_logic_vector(n-1 downto 0));
20   end register1;
```

```
21
22  architecture arch of register1 is
23  signal qi: std_logic_vector(n-1 downto 0);
24  begin
25    q <= qi;
26    process(clk,rst,en)
27    begin
28      if rst = '0' then
29        qi <= initial;
30      elsif (rising_edge(clk)) then
31          if en = '1' then
32            qi <= d;
33          else
34          qi <= qi;
35          end if;
36      end if;
37    end process;
38  end arch;
```

Listing 2.8   VHDL description of a 32 bit register

Listing 2.9 shows the VHDL description of the counter that controls the iterations of Lorenz chaotic oscillator, it can count for n bit, and in the description $n = 5$. This block will generate a Logic "1" when the count reaches n.

```
1   ----------------------------------------------------------------
2   ---Project: Ascending Counter 0 - n                          ---
3   ---Instructions: Assign the inputs correctly.                ---
4   ---    Inputs: clk - according to FPGA Model, example 50 MHz ---
5   ---            rst - put a button/switch of FPGA             ---
6   ---    Output: load - Gives '1' when the count reaches n     ---
7   ----------------------------------------------------------------
8   library ieee;
9   use ieee.std_logic_1164.all;
10  use ieee.numeric_std.all;
11
12  entity counter is
13    generic(n: integer:= 5);
14    port (
15    clk,rst :in std_logic;
16    load  :out std_logic:='0');
17  end counter;
18
19  architecture arch of counter is
20  begin
21    secuencial: process(rst,clk)
22    variable count: integer:=0;
23    begin
24      if rst = '0' then
25        count := 0;
26      elsif (rising_edge(clk)) then
27            if count = n then
28          load <= '1';
29              count := 0;
30            else
31            load <= '0';
32              count := count + 1;
33            end if;
34      end if;
35    end process;
36  end arch;
```

Listing 2.9   VHDL description of a counter

Using the VHDL descriptions listed above, one can describe the whole Lorenz chaotic oscillator. Listing 2.10 shows the entity that has two inputs (clk and rst) and three output ports to observe the state variables x, y, z, and architecture has a part to declare signals that are used into the VHDL description but that they are not related to the input or output ports. This block description calls to the VDHL blocks described above for the ADDER, SUBTRACTOR, MULTIPLIER, SCM, COUNTER, and REGISTER. They are called using U01, U02, and so on to describe the equations of Lorenz chaotic oscillator.

```
1    -----Project: Lorenz chaotic oscillator simulated with
2    ---- Forward-Euler method and h=0.01              ----
3    ----Instructions: Assign the inputs and outputs correctly.          ----
4    ----     Inputs: clk - according to FPGA Model, example 50 MHz       ----
5    ----             rst - button/switch of FPGA            ----
6    ----         Outputs: xf, yf and zf - Send to a DAC     ----
7    ----                  to view in an oscilloscope
8    ----IMPORTANT: Assign initial conditions to xi,yi,zi signals---
9    ------------------------------------------------------------------
10   library ieee;
11   use ieee.std_logic_1164.all;
12   use ieee.numeric_std.all;
13
14   entity lorenz_oscillator is
15     generic(n: integer:= 32);
16     port (
17     clk,rst : in std_logic;
18     xf,yf,zf: out std_logic_vector(n-1 downto 0)
19     );
20   end lorenz_oscillator;
21
22   architecture oscillator of lorenz_oscillator is
23   -- Signals declaration (All are used to interconnect the blocks)
24   signal xi,yi,zi     :std_logic_vector(n-1 downto 0):=x"0001999A";
25   -- Initial Conditions
26   signal x,y,z              :std_logic_vector(n-1 downto 0);
27   signal kx1,ky1,kz1        :std_logic_vector(n-1 downto 0);
28   signal funx,funy,funz     :std_logic_vector(n-1 downto 0);
29   signal ax1                :std_logic_vector(n-1 downto 0);
30   signal ay1,ay2,ay3        :std_logic_vector(n-1 downto 0);
31   signal az1,az2            :std_logic_vector(n-1 downto 0);
32   signal en                 :std_logic:='0';
33   -- Mapping the chaotic system
34   begin
35   -- Description of the equation of the derivate of variable x
36   -- dx = s*(y-x)
37   U01: entity work.substractor2(arch) generic map(n)
38                                       port map(clk,rst,yi,xi,ax1);
39   U02: entity work.scm_10(arch)
40                                       port map(clk,rst,ax1,funx);
41   U03: entity work.scm_h(arch)
42                                       port map(clk,rst,funx,kx1);
43   U04: entity work.adder2(arch)       generic map(n)
44                                       port map(clk,rst,xi,kx1,x);
45   -- Description of the equation of the derivate of variable y
46   -- dy = -x*z+r*x-y;
47   U05: entity work.scm_28(arch)
48                                       port map(clk,rst,xi,ay1);
49   U06: entity work.multiplier2(arch)  generic map(n)
50                                       port map(clk,rst,xi,zi,ay2);
51   U07: entity work.substractor2(arch) generic map(n)
52                                       port map(clk,rst,ay1,ay2,ay3);
53   U08: entity work.substractor2(arch) generic map(n)
54                                       port map(clk,rst,ay3,yi,funy);
55   U09: entity work.scm_h(arch)
```

```
56                                      port map(clk,rst,funy,ky1);
57  U10: entity work.adder2(arch)       generic map(n)
58                                      port map(clk,rst,yi,ky1,y);
59  -- Description of the equation of the derivate of variable z
60  -- dz = x*y-b*z
61  U11: entity work.multiplier2(arch)  generic map(n)
62                                      port map(clk,rst,xi,yi,az1);
63  U12: entity work.scm_8_3(arch)      
64                                      port map(clk,rst,zi,az2);
65  U13: entity work.substractor2(arch) generic map(n)
66                                      port map(clk,rst,az1,az2,funz);
67  U14: entity work.scm_h(arch)        
68                                      port map(clk,rst,funz,kz1);
69  U15: entity work.adder2(arch)       generic map(n)
70                                      port map(clk,rst,zi,kz1,z);
71  -- Description of the counter and registers
72  U16: entity work.counter(arch)      generic map(5)
73                                      port map(clk,rst,en);
74  U17: entity work.register1(arch)    generic map(n,xi)
75                                      port map(clk,rst,en,x,xi);
76  U18: entity work.register1(arch)    generic map(n,yi)
77                                      port map(clk,rst,en,y,yi);
78  U19: entity work.register1(arch)    generic map(n,zi)
79                                      port map(clk,rst,en,z,zi);
80  -- Assign the outputs through signals
81  xf <= xi;   yf <= yi;   zf <= zi;
82  end oscillator;
```

Listing 2.10 VHDL description of Lorenz chaotic oscillator

2.5 Experimental Observation of Chaotic Attractors Truncating the Fixed-Point Representation

The experimental observation of the chaotic attractors of Lorenz circuit is shown in Figs. 2.19 and 2.20. In the former figure one can observe the phase-space portraits of the experimental attractors of the FPGA-based Lorenz chaotic oscillator applying Forward Euler, and in the other figure applying the fourth-order Runge-Kutta method, respectively. The experimental observations were done using a Teledyne Lecroy Oscilloscope 2.5 GHz, using a digital-to-analog (DAC DAS1612VA) converter of 16 bit, with a conversion speed of 200 kSamples/s, and the signal measurement was conditioned as shown in Fig. 2.21. As one sees, the state variables provided by the FPGA are digital words of 32 bit (x_{n+1}, y_{n+1}, z_{n+1}). However, to observe those data in the oscilloscope, the DAC can process only 16 bit, and they must be the bits that represent the dominant dynamical behavior. By analyzing the binary representation of the real numbers, one can find the 16 bit from the state variables, which maintain the chaotic dynamics, and then one can connect each state variable from the FPGA to the DAC, as sketched in Fig. 2.21, which provides two outputs that are multiplexed to observe two state variables in the oscilloscope.

An example on the resolution that is generated by performing the truncation is given in Table 2.3. The state variables provided by the FPGA have length [31:0

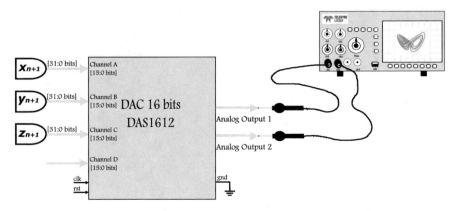

Fig. 2.21 Block description of the DAC that uses 16 bit and the state variables are truncated from 32 to 16 bit to observe the experimental attractors in a Teledyne Lecroy oscilloscope

Table 2.3 Example of truncation of 32 bit represented in fixed-point notation, to a 16 bit word for the real number 50.1251

Real number	Fixed-point format	Sign bit	Integer part	Fractional part	Truncated number
50.1251	12.20	1 Bit	11 Bits	20 Bits	50.1251
		0	00000110010	.00100000000001101001	
	7.9	1 Bit	6 Bits	9 Bits	50.1250
		0	110010	.001000000	

bits], and they must be truncated to be processed through the DAC that can receive data of [15:0 bits]. In this manner, because one knows a priori that the ranges of the values of the state variables are between [−50,50], then one must find the digital representation of 16 bit that guarantees the best resolution. Therefore, the first column in Table 2.3 shows the real number being used to analyze the effect of truncating from 32 bit to 16 bit, and it is the real number 50.1251. The second column shows the fixed-point format using 32 bit having representation of 12.20, and then truncating to 16 bit having representation of 7.9. The third, fourth, and fifth columns show the meaning of the bits in each case having 32 and 16 bit, but maintaining the representation of the real number 50.1251. In particular, one can see that the fractional part using 32 bit is truncated to 9 bit and the less significative ones are eliminated, thus generating loss of information. The sixth column shows the real number that can be represented using fixed-point format with 16 bit, thus having the real number equal to 50.1250. This is a small loss of information and for this reason one can observe the experimental attractors in the oscilloscope.

If the truncation is not performed in the right way, the attractors cannot be observed. On one hand, truncating the 16 less significative bits from the original 32

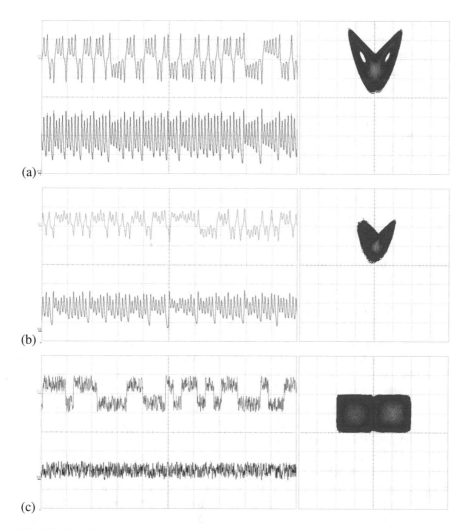

Fig. 2.22 Experimental observation of the chaotic attractor of the FPGA-based Lorenz oscillator showing the state variables x and z, and truncating the original 32 bit word to 16 bit as shown in Fig. 2.21. From the [31:0] bits in the fixed-point format 12.20: (**a**) using the bits 31, 25–11 with axes 2 V/Div, (**b**) using the bits 31–16 with axes 100 mV/Div, and (**c**) using the bits 31, 14–0 with axes 10 V/Div

bits, one can have loss of information in the fractional numbers because the integer part of the representation will have the most significant bits useless. On the other hand, truncating the 16 most significative bits, the fractional part may be perfectly reconstructed but not the integer part. This is sketched in Fig. 2.22, where it is very clear that the best observation of the chaotic attractor of the FPGA-based Lorenz oscillator, when truncating the original 32 bit word to 16 bit, is shown in Fig. 2.22a.

This case is the one listed in Table 2.3, which shows that the 32 bit notation [31:0] in the fixed-point format 12.20 can be truncated to use the sign bit [31], the six less significative bits of the integer part [25:20], and the nine most significative bits of the fractional part [19:11]. Figure 2.22b shows the attractor when using the most significative bits [31:16], the loss of information is evident, and the attractor is not appreciated as in Fig. 2.22a. In fact, the sign is still represented by one bit [31], and the integer part is also still represented using the 11 bits [30:20]; however, the fractional part will be represented only by the remaining 4 bits [19:15], so that the resolution is lost. The worst case is when using the sign bit [31] and the 15 less significative bits [14:0], where the fractional part in the original fixed-point format is 12.20, and then the integer part cannot be represented and the fractional part will only be represented by 15 bits of the 20 bits that are required in the 32 bits word. This case is shown in Fig. 2.22c, thus confirming that the truncation operation is not a trivial task and one must analyze the computer arithmetic that is used. A similar case is presented when implementing the fractional-order chaotic oscillators, as shown in the next chapters.

The FPGA used resources for observing the chaotic attractors of Lorenz oscillator are given in Table 2.4, using Cyclone IV GX EP4CGX150DF31C7 from ALTERA. The one-step methods detailed in Chap. 1, the Forward Euler, and the fourth-order Runge-Kutta are used, which require different number of resources and then each numerical method provides different operating frequency and latency. As one can predict, the Forward Euler can provide the highest frequency because it requires less resources, but recall that the time-step h must be selected to generate the lower numerical error. The data listed in Table 2.4 were obtained from the FPGA synthesizer Quartus 13.0. The maximum frequency is multiplied by the number of clock cycles that are required to process the data from the input to the output. This is equivalent to evaluating the processing speed or latency, which is listed in the last column of the table and given in nanoseconds (ns). Comparing both numerical methods, one can observe that the fourth-order Runge-Kutta method uses approximately four times more resources than Forward-Euler, but it has more exactness when performing the integration operation. This is a trade-off that involves thinking on reducing hardware resources or enhancing the integration accuracy of the equations modeling the chaotic oscillator.

It is worth mentioning that the FPGA-based implementation of chaotic oscillators can find several engineering applications like random number generation and secure communications [13]. Other old applications can be enhanced using FPGAs, for example: encoding of the information for private communication [14, 15], modeling of natural processes [16], modeling of chemical and biological systems [17], and so on. Several examples are given in Chap. 6, using fractional-order chaotic oscillators that are implemented on FPGAs.

Table 2.4 FPGA used resources for Lorenz chaotic oscillator when applying two one-step numerical methods and using the FPGA Cyclone IV GX EP4CGX150DF31C7

Chaotic system	Numerical method	Logic elements	Registers	9*9 bits multiplier	Maximum frequency (MHz)	Cycles	Iteration latency (ns)
Lorenz oscillator	Forward-Euler	1411	730	25	93.5	5	53.47
	Fourth-order Runge-Kutta	5652	3420	100	85.56	20	233.7

References

1. R. Trejo-Guerra, E. Tlelo-Cuautle, V.H. Carbajal-Gómez, G. Rodriguez-Gomez, A survey on the integrated design of chaotic oscillators. Appl. Math. Comput. **219**(10), 5113–5122 (2013)
2. V.H. Carbajal-Gomez, E. Tlelo-Cuautle, J.M. Muñoz-Pacheco, L.G. de la Fraga, C. Sanchez-Lopez, F.V. Fernandez-Fernandez, Optimization and CMOS design of chaotic oscillators robust to PVT variations. Integration **65**, 32–42 (2018)
3. W.M. Ahmad, J.C. Sprott, Chaos in fractional-order autonomous nonlinear systems. Chaos Solitons Fractals **16**(2), 339–351 (2003)
4. C. Muñiz-Montero, L.V. García-Jiménez, L.A. Sánchez-Gaspariano, C. Sánchez-López, V.R. González-Díaz, E. Tlelo-Cuautle, New alternatives for analog implementation of fractional-order integrators, differentiators and PID controllers based on integer-order integrators. Nonlinear Dynam. **90**(1), 241–256 (2017)
5. J.-M. Garcia-Ortega, E. Tlelo-Cuautle, C. Sanchez-Lopez, Design of current-mode Gm-C filters from the transformation of opamp-RC filters. J. Appl. Sci. **7**(9), 1321–1326 (2007)
6. C. Muñiz-Montero, L.A. Sánchez-Gaspariano, C. Sánchez-López, V.R. González-Díaz, E. Tlelo-Cuautle, On the electronic realizations of fractional-order phase-lead-lag compensators with OpAmps and FPAAs, in *Fractional Order Control and Synchronization of Chaotic Systems* (Springer, Berlin, 2017), pp. 131–164
7. F. Rahma et al., Analog programmable electronic circuit-based chaotic Lorenz system. Basrah J. Eng. Sci. **14**(1), 39–47 (2014)
8. Anadigm, Dynamically Reconfigurable dpASP, 3rd Generation, AN231E04 Datasheet Rev 1.2 (2014). www.anadigm.com
9. Anadigmdesigner2 user manual (2014). www.anadigm.com
10. O. Guillén-Fernández, A. Meléndez-Cano, E. Tlelo-Cuautle, J.C. Núñez-Pérez, J. de Jesus Rangel-Magdaleno, On the synchronization techniques of chaotic oscillators and their FPGA-based implementation for secure image transmission. PloS One **14**(2), e0209618 (2019)
11. E. Tlelo-Cuautle, J.J. Rangel-Magdaleno, A.D. Pano-Azucena, P.J. Obeso-Rodelo, J.C. Nuñez-Perez, FPGA realization of multi-scroll chaotic oscillators. Commun. Nonlinear Sci. Numer. Simul. **27**(1–3), 66–80 (2015)
12. E. Tlelo-Cuautle, A.D. Pano-Azucena, J.J. Rangel-Magdaleno, V.H. Carbajal-Gomez, G. Rodriguez-Gomez, Generating a 50-scroll chaotic attractor at 66 MHz by using FPGAs. Nonlinear Dynam. **85**(4), 2143–2157 (2016)
13. E. Tlelo-Cuautle, L.G. de la Fraga, J. Rangel-Magdaleno, *Engineering Applications of FPGAs* (Springer, Berlin, 2016)
14. G. Cai, Z. Tan, Chaos synchronization of a new chaotic system via nonlinear control. J. Uncertain Syst. **1**(3), 235–240 (2007)
15. A. Kiani-B, K. Fallahi, N. Pariz, H. Leung, A chaotic secure communication scheme using fractional chaotic systems based on an extended fractional Kalman filter. Commun. Nonlinear Sci. Numer. Simul. **14**(3), 863–879 (2009)
16. E.N. Lorenz, Deterministic nonperiodic flow. J. Atmos. Sci. **20**(2), 130–141 (1963)
17. R.J. Field et al., *Chaos in Chemistry and Biochemistry* (World Scientific, Singapore, 1993)

Chapter 3
Characterization and Optimization of Fractional-Order Chaotic Systems

3.1 Fractional-Order Chaotic Oscillators

The integer-order chaotic oscillators described in Chap. 1 can be transformed into fractional-order ones. In fact, as mentioned in the first chapter of this book, the integer-order system can be considered as a particular case of the fractional-order one, which is more general. Among all the integer-order chaotic oscillators, three of them are considered herein to be optimized in their positive Lyapunov exponent and Kaplan–Yorke dimension (D_{KY}) applying differential evolution (DE) and particle swarm optimization (PSO) algorithms.

The fractional-order Lorenz's chaotic oscillator [1], which is the general version of the integer-order one given in (1.1), is described by the equations given in (3.1). In 1999, Chen introduced another three-dimensional autonomous system [2], which is topologically different to Lorenz's chaotic oscillator and which has a chaotic attractor as well. Chen's chaotic oscillator given in (1.13), and its fractional-order description is given in (3.2). The third chaotic oscillator that is taken as case of study in this chapter to be optimized is the heart-shape given in (1.16), which is based in the Lorenz's attractor and introduced in [3]. The fractional-order heart-shape chaotic oscillator presents multiple shifts when adding a nonlinear function $G(z)$, with z as a state variable, and also by adding parameter m, as shown in the equations given in (3.3). The nonlinear function of ladder type described by $G(z)$ can be approached by a step-piecewise function as shown in (3.4), where d_i is an additive coefficient, S_i is a limiting coefficient, and the number of scrolls that can be generated is established by $2N$.

$$
\begin{aligned}
{}_0D_t^{q_1}x(t) &= \sigma(y(t) - x(t)), \\
{}_0D_t^{q_2}y(t) &= x(t)(\rho - z(t)) - y(t), \\
{}_0D_t^{q_3}z(t) &= x(t)y(t) - \beta z(t),
\end{aligned}
\tag{3.1}
$$

© Springer Nature Switzerland AG 2020

E. Tlelo-Cuautle et al., *Analog/Digital Implementation of Fractional Order Chaotic Circuits and Applications*, https://doi.org/10.1007/978-3-030-31250-3_3

$$\begin{aligned}
{}_0D_t^{q_1}x(t) &= a(y(t) - x(t)), \\
{}_0D_t^{q_2}y(t) &= (c-a)x(t) - x(t)z(t) + cy(t), \\
{}_0D_t^{q_3}z(t) &= x(t)y(t) - bz(t),
\end{aligned} \tag{3.2}$$

$$\begin{aligned}
{}_0D_t^{q_1}x(t) &= y(t) - x(t), \\
{}_0D_t^{q_2}y(t) &= sign(x(t))[1 - mz(t) + (Gz(t))], \\
{}_0D_t^{q_3}z(t) &= |x(t)| - rz(t)
\end{aligned} \tag{3.3}$$

$$G(Z) = \begin{cases}
0 & Z < S_0 \\
d_1 & S_0 \le Z < S_1 \\
\vdots & \vdots \\
d_{N-1} & Z \ge S_{N-1}
\end{cases} \tag{3.4}$$

The parameters or coefficients and the nonlinear functions of the fractional-order chaotic oscillators described by (3.1)–(3.3) can be varied in order to find the best values of their Lyapunov exponents and D_{KY}. This should be a challenge because a huge number of combinations may arise, and therefore, metaheuristics can be a good option to perform the optimization process. In this manner, DE and PSO algorithms are applied to vary the design variables in a range specified by the user. For instance, the three fractional-order chaotic oscillators given above can be simulated using the design variables given in the literature, which are listed in Table 3.1. The table also gives the fractional-order of the derivatives, the positive Lyapunov exponent (LE+) value, and the associated D_{KY}.

In fractional-order chaotic oscillators, one can find the equilibrium points and eigenvalues, as shown in Chap. 1, where the analysis was performed for integer-order ones. However, the same analysis applies for fractional-order chaotic oscillators, with the goal of determining unstable regions. After one evaluates the equilibrium points by setting the derivatives equal to zero, the eigenvalues can be calculated for each equilibrium point and using the Jacobian matrix. In this manner, the equilibrium points of (3.1)–(3.3) and their associated Jacobian matrixes are given in Table 3.2. Each fractional-order chaotic oscillator will have three eigenvalues for each equilibrium point because all of them have three state variables. For more complex fractional-order chaotic oscillators, having more equilibrium points or having more than three state variables, the evaluation of the eigenvalues may require numerical methods, and then one can apply Cardano's method [4].

Table 3.1 LE$_+$ and D_{KY} of Eqs. (3.1)–(3.3) reported in the literature

System	Design variables	Fractional order	LE$_+$	D_{KY}
Lorenz	$\sigma = 10, \rho = 28, \beta = 8/3$	$q_1 = q_2 = q_3 = 0.994$	0.8500	2.0900
Chen	$a = 35, b = 3, c = 28$	$q_1 = q_2 = q_3 = 0.6$	0.0034	2.0791
Heart shape	$m = 2, r = 0.05$	$q_1 = q_2 = q_3 = 0.95$	0.4060	–

Table 3.2 Equilibrium points and Jacobian matrix of each fractional-order chaotic oscillator given in (3.1)–(3.3)

System	Jacobian	Equilibrium points
Lorenz	$\begin{bmatrix} -\sigma & \sigma & 0 \\ \rho - z & -1 & -x \\ y & x & \beta \end{bmatrix}$	$\left(\pm\sqrt{\beta(\rho - 1)}, \pm\sqrt{\beta(\rho - 1)}, \rho - 1\right)$
Chen	$\begin{bmatrix} -a & a & 0 \\ c - a - z & c & -x \\ y & x & -\beta \end{bmatrix}$	$\left(\pm\sqrt{b(2c - a)}, \pm\sqrt{b(2c - a)}, 2c - a\right)$
Heart shape	$\begin{bmatrix} -1 & 1 & 0 \\ 0 & 0 & \pm m \\ \pm 1 & 0 & -r \end{bmatrix}$	$\left(\pm\dfrac{z_i}{2}, \pm\dfrac{z_i}{2}, z_i\right)$

Using the values of the design variables listed in Table 3.1, then one can perform numerical simulations to observe the attractors. As shown in Chap. 1, one can apply approximations like Grüwald–Letnikov, Adams–Bashforth–Moulton, Caputo's definition, FDE12 available into MatLab^{TM} [5, 6], and so on. In this manner, applying FDE12, the phase-space portraits of the fractional-order chaotic attractors are shown in Fig. 3.1, where it can be appreciated that the step-size h is different for the three fractional-order chaotic oscillators. In fact, selecting h can be time-consuming when it is done in a trial and error way, and selecting a bad h can lead to artificial chaos [7]. That way, h can be estimated from all the eigenvalues that are computed for each equilibrium point. Therefore, one must be interested in finding imaginary eigenvalues because they originate oscillatory behavior. The imaginary eigenvalues have the form $\lambda_i = \mathrm{Re}_i + \mathrm{Im}(\omega_i)$, where subindex i denotes the λ_i-th eigenvalue having real Re_i and imaginary $\mathrm{Im}(\omega_i)$ parts. This task can be automated within an optimization process, so that the algorithm should search for a special class of eigenvalues that are computed from all the equilibrium points. Another challenge arises when the chaotic attractors are hidden, because one cannot evaluate the equilibrium points that can be infinite or difficult to evaluate by analytical methods, as discussed in Chap. 1.

3.2 Computing Lyapunov Exponents Applying Wolf's Method and TISEAN

3.2.1 Evaluating Lyapunov Exponents by Wolf's Method

In 1985 Wolf [8] introduced an algorithm focused on finding the largest Lyapunov exponent based on the time series of data for chaotic systems. The direct calculation of the first Lyapunov exponent as per Wolf's method is based on keeping the track

Fig. 3.1 Fractional-order chaotic attractors of the oscillators described by (3.1)–(3.3). The design parameters were taken from Table 3.1, and simulated with h and initial conditions $(x(0), y(0), z(0))$ equal to: (**a**) $h = 0.003$ and $(0.1,0.1,0.1)$, (**b**) $h = 0.001$ and $(-9,-5,14)$, and (**c**) $h = 0.005$ and $(0.5,1,0.5)$

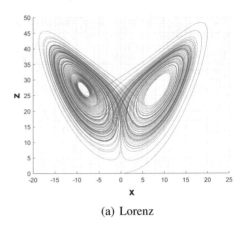

(a) Lorenz

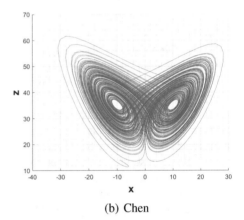

(b) Chen

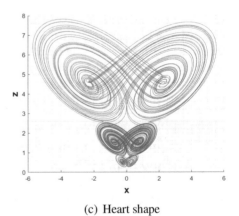

(c) Heart shape

of the exponentially divergent adjoining trajectories. The degree of the trajectory divergence is evaluated at regular intervals [9]. The main function of Wolf's method is the identification of a pair of points and the measurement of the distance of their associated reconstructed vectors as the system evolves in time by the time steps of duration t_s seconds each. If this distance is large enough (more specifically this distance should be greater than a minimum distance D_{min} and less than a maximum distance D_{max}), the method searches for a new point to substitute one of the points of the current pair.

Wolf's method requires a technique for simultaneously estimating all n Lyapunov exponents, and it relies on the orthonormalization of n vectors. Such a technique is known as Gram–Schmidt orthonormalization, and it is a procedure that can be described as follows: Let $\{\delta x_1, \ldots, \delta x_n\}$ be a set of linearly independent n-vectors in IR^n. Gram–Schmidt orthonormalization generates an orthonormal set $\{u_1, \ldots, u_n\}$ of n-vectors with the property that $\{u_1, \ldots, u_i\}$ spans the same subspace as $\{\delta x_1, \ldots, \delta x_i\}$ for $i = 1, \ldots, n$. The evaluations can be performed as follows:

$$v_1 = \delta x_1$$

$$u_1 = v_1 / \|v_1\|$$

$$v_2 = \delta x_2 - \langle \delta x_2, u_1 \rangle u_1 \quad u_2 = v_2 / \|v_2\| \qquad (3.5)$$

$$\vdots$$

$$v_n = \delta x_n - \langle \delta x_n, u_1 \rangle u_1 - \ldots - \langle \delta x_n, u_{n-1} \rangle u_n = v_n / \|v_n\|$$

A simple calculation shows that for any $m \leq n$, $\langle u_m, u_i \rangle = 0$ for $1 \leq i < m$. It follows that $\{u_1, \ldots, u_n\}$ is an orthogonal set. Furthermore, since u_i is a linear combination of $\delta x_1, \ldots, \delta x_{i-1}$, it is clear that $\{u_1, \ldots, u_i\}$ and $\{\delta x_1, \ldots, \delta x_i\}$ span the same subspace for $i = 1, \ldots, n$.

Just as in the two-dimensional case, the area of the parallelepiped spanned by $\{\delta x_1, \ldots, \delta x_i\}$ is

$$Volume\{\delta x_1, \ldots, \delta x_i\} = \|v_1\| \ldots \|v_i\| \qquad (3.6)$$

for $i = 1, \ldots, n$.

As sketched in Algorithm 1 for evaluating the positive Lyapunov exponent LE+, it is necessary to solve the extended system of each of the systems of fractional-order, which implies solving the Jacobian matrix in step number 7 and that is shown for the three systems in Table 3.2.

The numerical integrations required by the evaluation of LE+ for fractional-order chaotic oscillators can be performed applying Grünwald–Letnikov method, the predictor-corrector Adams–Bashforth–Moulton method [10], and so on. However,

---

Algorithm 1 Algorithm for evaluating LE+

Require: $neq, x\_start, [t\_start, t\_end], h\_norm$, and $n\_i(t) \leftarrow (t\_end - t\_start)/h\_norm$
    neq = number of equations,
    $x\_start =_{neq}$ initial conditions of (3.1)–(3.3),
    $[t\_start, t\_end]$ = time span,
    h_norm = normalization of step-size for Gram-Schmidt,
    $n\_i(t) \leftarrow (t\_end - t\_start)/h\_norm$ = iterations number,
 1: **begin**
 2: **for** $i \leftarrow neq + 1$ **to** $neq(neq + 1)$ **do**
 3:     $x(i) = 0.1$
 4: **end for**
 5: $t \leftarrow t\_start$
 6: **for** $i \leftarrow 1$ **to** n_it **do**
 7:     $x \leftarrow$ integration of systems (3.1), (3.2), (3.3)
 8:     $t \leftarrow t + h\_norm$
 9:     $zn(1), \ldots, zn(neq) \leftarrow$ Gram-Schmidt Procedure
10:     $s(1) \leftarrow 0$
11:     **for** $k \leftarrow 1$ **to** ne **do**
12:         $s(k) \leftarrow s(k) + log(zn(k))$ vector magnitudes
13:         $LE(k) \leftarrow s(k)/(t - t\_start)$
14:     **end for**
15: **end for**
16: **return** $LE = 0$

---

due to the requirement of manipulating huge memory blocks for fractional-order chaotic oscillators, one must adapt the approximations, for example as done in [11], and linking Wolf's method described in Algorithm 1. A fast approximation can be performed applying the method introduced in [12, 13].

3.2.2   The Lyapunov Spectrum Provided by TISEAN

The Lyapunov exponents of a fractional-order chaotic oscillator can be determined by using the chaotic time series generated from the numerical simulation when using TISEAN (TIme SEries ANalysis). In this case, the exactness will depend on the numerical approximation and the length of the data in the time series. For instance, a reliable characterization requires that the independence of embedding parameters and the exponential law for the growth of distances be checked as suggested in [14, 15].

Let us consider the representation of the chaotic time series data as a trajectory in the embedding space, and assume that one can observe a very close return $s_{n'}$ to a previously visited point s_n. Then one can consider the distance $\Delta_0 = s_n - s_{n'}$ as a small perturbation, which should grow exponentially in time. Its future can be read from the time series $\Delta_l = s_{n+l} - s_{n'+l}$. If one finds that $|\Delta_l| \approx \Delta_0 e^{\lambda l}$, then λ is (with probability one) the maximum Lyapunov exponent. In practice, there will be fluctuations because of many effects, which are discussed in detail in [14]. Based on this understanding, one can derive a robust consistent and unbiased estimator for the maximum Lyapunov exponent. One can compute (3.7), in which $S(\epsilon, m, t)$ exhibits a linear increase with identical slope for all m larger than some m_0 and for a reasonable range of ϵ, then this slope can be taken as an estimation of the maximum exponent λ_1 [16].

$$S(\epsilon, m, t) = \left\langle \ln \left(\frac{1}{|U_n|} \sum_{s_{n'} \in U_n} |s_{n+t} - s_{n'+t}| \right) \right\rangle \qquad (3.7)$$

This mathematical approximation of the Lyapunov exponents is implemented in TISEAN under the routine lyapspec, and the parameters are calibrated as shown in Table 3.3. It can be good if the chaotic time series data has a length of 250,000 samples for the three state variables x, y, z of the fractional-order chaotic oscillators. If the transient behavior is not deleted previously, TISEAN can eliminate a desired number of lines with the option -x, as shown in the second row of Table 3.3. Also, one can use any number of columns that are associated with the discrete time and values of the chaotic time series.

The output provided by TISEAN consists of $d * m + 1$ columns. The first one shows the actual iteration, and the next $d * m$ ones show the estimations of the Lyapunov exponents in decreasing order. The last lines show the average forecast error(s) of the local linear model, the average neighborhood size used for fitting the model and the very last line the estimated Kaplan–Yorke dimension. This information can be extracted and then this process can be automated and linked to an optimization loop as described below.

Table 3.3 Input parameters to estimate the Lyapunov spectrum of a time series using Lyapspec in TISEAN

Option	Description	Value
-l	Number of points to use	250,000
-x	Number of lines to be ignored	10,000
-c	Column to be read	1,2,3
-m	No. of components, embedding dimension	1,3
-d	Delay for the delay vectors	1
-r	Initial size of the neighborhoods	0.1
-f	Factor to increase the size of the neighborhood	1.2
-k	Number of neighbors to use	30
-n	Number of iterations	1000

3.3 Optimizing Fractional-Order Chaotic Oscillators Applying DE and PSO Algorithms

As mentioned above, chaotic systems can be optimized in order to provide better characteristics like high LE+ and high D_{KY} [17]. This can be done by applying metaheuristics because the design variables have large search spaces and thus require extensive computing time. A clear example is when trying to optimize the Lorenz chaotic oscillator, which has three design parameters: σ, ρ, and β, and they can have integer and fractional values. Therefore, if the design parameters are varied using two integer digits to represent from 0 to 99 one has 10^2 combinations, and if those integer numbers include four fractional numbers to represent from 0.0001 to 99.9999 then one has 10^6 combinations. In this manner, having three design variables (σ, ρ, and β), then the number of combinations becomes $10^6 \times 10^6 \times 10^6 = 10^{18}$. As one can see, simulating this number of cases can be unreachable in a couple of years and the most important thing is that not all the combination cases will generate chaotic behavior. Another justification of applying metaheuristics is that an algorithm is used to evaluate D_{KY}, which is not based on derivatives. The D_{KY} requires evaluating the Lyapunov exponents of a chaotic oscillator, for which several methods have been published [18–20].

As already mentioned in the previous chapters, the fractional-order chaotic oscillators are simulated using methods different to the ones applied to integer-order ones. For fractional-order systems one can apply the different approximations known as: Grünwald–Letnikov [21], given in (3.8), where $[x]$ means the integer part of x and q and t are the bounds of operation for $_aD_t^q f(t)$; Riemmann–Liouville [22], given in (3.9), where $\Gamma(\cdot)$ is the well-known Euler's Gamma function; Caputo [23], given in (3.10); and so on. More elaborated methods based on predicting and correcting are for example, the Adams–Bashforth–Moulton of type Predictor-Corrector [11, 24], given in (3.11) and (3.12). In the time domain methods like Grünwald–Letnikov can reduce hardware resources when using the short memory principle described in Chap. 1. In such a case L_m denotes the length of memory, $t_k = kh$, h is associated with the time-step, and $(-1)^j \binom{q}{j}$ are the binomial coefficients C_j^q ($j = 0, 1, \ldots$).

$$_aD_t^q f(t) = \lim_{h \to 0} \frac{1}{h^q} \sum_{j=0}^{\left[\frac{t-a}{h}\right]} (-1)^j \binom{q}{j} f(t - jh) \tag{3.8}$$

$$_0D_t^q f(t) = \frac{1}{\Gamma(n-q)} \frac{d^n}{dt^n} \int_0^t \frac{f(\tau)}{(t-\tau)^{q-n+1}} d\tau \tag{3.9}$$

$$_aD_t^q f(t) = \frac{1}{\Gamma(n-q)} \int_a^t \frac{f^{(n)}(\tau)}{(t-\tau)^{q-n+1}} d\tau, \qquad \text{for } n - 1 < q < n. \tag{3.10}$$

$$y_h^p(t_n + 1) = \sum_{k=0}^{m-1} \frac{t_{n+1}^k}{k} y_0^{(k!)} + \frac{1}{\Gamma(q)} \sum_{k=0}^{n} b_{j,n+1} f(t_j, y_n(t_j)) \tag{3.11}$$

$$y_h(t_{n+1}) = \sum_{k=0}^{m-1} \frac{t_{n+1}^k}{k!} y_0^{(k)} + \frac{h^q}{\Gamma(\alpha + 2)} f\left(t_{n+1}, y_h^p(t_{n+1})\right)$$

$$+ \frac{h^q}{\Gamma(\alpha + 2)} \sum_{j=0}^{n} a_{j,n+1} f(t_j, y_n(t_j)), \tag{3.12}$$

Apart from applying the definitions for simulating fractional-order chaotic oscillators, it may be necessary to consider some conditions to guarantee that the dynamical system has a chaotic behavior. For instance, at least one eigenvalue must be in the unstable region so that the value of the fractional-order of the derivatives associated with q will be the one to evaluate the whole order of the system, and for a commensurate system $q_1 = q_2 = q_3 = q$. Therefore, one can use (3.13) to estimate the order higher than 2 for a fractional-order chaotic oscillator, where $Im((\lambda)$ and $Re(\lambda)$ denote the imaginary and real parts of an eigenvalue [25].

$$q > \frac{2}{\pi} arctan \frac{|Im((\lambda)|}{|Re(\lambda)|}, \tag{3.13}$$

For incommensurate fractional-order chaotic oscillators, in which $q_1 \neq q_2 \neq q_3$, a necessary condition to guarantee chaotic behavior is equivalent to evaluate (3.14), where λ_i are the eigenvalues of (3.15)

$$\left(\frac{\pi}{2M}\right) - \min_i \{|arg(\lambda_i)|\} \geq 0 \tag{3.14}$$

$$det\left(diag\left(\begin{bmatrix} \lambda^{Mq_1} & \lambda^{Mq_2} & \lambda^{Mq_3} \end{bmatrix}\right) - \partial f / \partial x \mid_{x=x^*}\right) = 0 \qquad \forall x^* \in \Omega \tag{3.15}$$

Applying the definitions (3.8), (3.12), and (3.11), and considering (3.13) for commensurate systems or (3.14) for incommensurate ones, then it is possible to simulate the three fractional-order chaotic oscillators given in (3.1)–(3.3). The numerical simulation of the fractional-order chaotic oscillators provides time series of the three state variables x, y, z that are used to evaluate LE+ and D_{KY} applying Wolf's method [8] or TISEAN [16]. Any of the cases described above can be included into an optimization loop to maximize LE+ and D_{KY} applying metaheuristics like DE and PSO, which are described below.

3.3.1   Differential Evolution Algorithm

This optimization algorithm belongs to the focused-evolutionary family to solve mono-objective problems, and can be used in other multi-objective optimization algorithms. DE is an algorithm that begins from generating D-dimensional vectors randomly to create a population of the form: $x_{i,G}$, with $i = \{1, 2, 3, \ldots, N_p\}$, and G as the maximum number of generations. N_p is the number of vectors in the population, which evolves in G generations. The new vectors are generated in each generation by performing mutation and operations [26], which are evaluated using (3.16) and (3.17), respectively.

$$v_i^{g+1} \leftarrow x_c^g + F(x_a^g - x_b^g) \tag{3.16}$$

$$u_{ij}^{g+1} \leftarrow \begin{cases} v_{ij}^{g+1} & \text{if } [U(0, 1) \leq CR] \text{ or } rnbr(i) = j \\ x_{ij}^{g+1} & \text{if } [U(0, 1) > CR] \text{ and } rnbr(i) \neq j \end{cases} \tag{3.17}$$

In (3.16), a, b, and c are vectors that are randomly selected, g is the current generation, and $F \in [0, 2]$ is a mutant constant. In (3.17), $j = \{1, 2, \ldots, D\}$, $U(0, 1)$ is a function that returns a real number with uniform distribution and within the range $[0, 1)$, $CR \in [0, 1]$ is a crossover coefficient selected by the user, and $rnbr(i) \in [0, 1]$ is an index generated randomly. During the evaluation process, several operations are performed as follows: When evaluating the new generated vector (u_{ij}^{g+1}), if it is better than the previous vector (x_{ij}^{g+1}), then the new vector replaces the previous one and it will be part of the new population ($g+1$), otherwise, the new vector will be discarded. The DE algorithm stops when the maximum number of generations is reached (or other stop criterion is applied) and the values of optimization are retrieved. Algorithm 2 shows the pseudo-code of the DE algorithm that is detailed in [26], and it is adapted to maximize the positive Lyapunov exponent (LE+).

The Lyapunov exponents of the fractional-order chaotic oscillators can be evaluated applying TISEAN or Wolf's method, which are well-known methods. Using TISEAN, one should calibrate the parameters. For example, step 2 in Algorithm 2 requires evaluating LE+, in this case the parameters required by TISEAN can be calibrated according to Table 3.3. This software is online and available for free, and requires the time series that are generated by the approximations described above.

3.3.2   Particle Swarm Optimization Algorithm

The PSO algorithm avoids performing a selection process as the evolutionary algorithms do. In PSO, all the population members survive in the whole optimization process. Basically, it updates the position and the velocity of the particles that follow

---

Algorithm 2 Maximizing LE+ applying differential evolution

Require: D, G, N_p, CR, F, and $func(\bullet)$
    D = random dimensional vectors, G = generations, N_p = vectors in the population
    CR and F are parameters of DE, and $func(\bullet)$ allocates the chaotic oscillator
  1: Initialize the population randomly $(\mathbf{x})$
  2: Evaluate fx (LE+) for each individual in the population with (3.8)–(3.10) or (3.11)
    This step takes the position values after performing the following operations:
    Obtain the equilibrium points and evaluate the Jacobian associated with the fractional-order
    chaotic oscillator to:
    Compute the eigenvalues to estimate h
    Verify the condition of the fractional derivative with (3.13) and (3.14)
    Evaluate LE+ applying TISEAN or Wolf's method [8]
  3: Save the LE+ results in *score*
  4: **for** $(counter = 1; counter \leq G; counter{+}{+})$ **do**
  5:   **for** $(i = 1; i \leq N_p; i{+}{+})$ **do**
  6:     Select three different indexes randomly (a, b, and c in (3.16))
  7:     **for** $(j = 1; j \leq D; j{+}{+})$ **do**
  8:       **if** $U(0, 1) < CR \parallel j = D$ **then**
  9:         $trial_j \leftarrow x_{aj} + F(x_{bj} - x_{cj})$
10:       **else**
11:         $trial_j \leftarrow x_{ij}$
12:       **end if**
13:     **end for**
14:     $fx \leftarrow func(trial)$
15:     **if** fx is better than $score_i$ **then**
16:       $score_i \leftarrow fx$
17:       $x_i \leftarrow trial$
18:     **end if**
19:   **end for**
20: **end for**
21: **return** x and $score = 0$

---

the particle with the best result. The particles are associated with $x_i \in \mathbb{R}^J$ vectors, which are randomly initialized. The vectors are viewed as particles in the space, and their behaviors are defined by two formulas associated with their velocity described by (3.18), and position described by (3.19).

$$v_{ij}^{t+1} \leftarrow \alpha v_{ij}^t + U(0, \beta)\left(p_{ij} - x_{ij}^t\right) + U(0, \beta)\left(g_j - x_{ij}^t\right) \tag{3.18}$$

$$x_{ij}^{t+1} \leftarrow x_{ij}^t + v_{ij}^{t+1} \tag{3.19}$$

In (3.18) and (3.19), i is the index of the particle, j is the dimension, p_i is the best position finding in i, and p_g is the best position obtained during the optimization. $\alpha \in \mathbb{R}$ is named inertial weight, $\beta \in \mathbb{R}$ is named acceleration constant, and $U(\bullet)$ is a generator of random numbers that are uniformly distributed.

PSO is based on evaluating a $f(x_i)$ function and comparing the results in the following way: If the last result is better than the i-th registered or previous saved results, this will be the new vector p_i. The global best value (g) is also compared, and if the last value is better than this, this is also replaced. This process is accomplished until it reaches the stop criterion. A feature of PSO is that during its execution, each particle moves around a centroid region determined by p_i and g, so that the particles pursue new positions to reach the best solution. Algorithm 3 shows the pseudo-code of PSO algorithm, which is detailed in [27], and it is adapted to maximize the positive Lyapunov exponent (LE+).

DE and PSO can also be adapted in a similar way to maximize the Kaplan–Yorke dimension D_{KY}, and both metaheuristics can be calibrated to provide better feasible solutions in less generations. One can also explore extending the capabilities of these algorithms to perform multi-objective optimization to maximize at the same time LE+ and D_{KY}, for instance.

3.4   Optimizing the Positive Lyapunov Exponent and Kaplan–Yorke Dimension Applying Metaheuristics

The Lyapunov exponents give the most characteristic description of the presence of a deterministic non-periodic flow. Therefore, Lyapunov exponents are asymptotic measures characterizing the average rate of growth (or shrinkage) of small perturbations to the solutions of a dynamical system [28]. Lyapunov exponents provide quantitative measures of response sensitivity of a dynamical system to small changes in initial conditions [29]. The number of Lyapunov exponents equals the number of state variables, and if at least one is positive, this is an indication of chaos [30]. That way, the three fractional-order chaotic oscillators described by (3.1)–(3.3) have three Lyapunov exponents, and their whole order must be higher than two. As those fractional-order chaotic oscillators have three Lyapunov exponents, in order to generate chaotic behavior, one Lyapunov exponent must be positive LE+, one must be zero (or very close to zero), and one must be negative. Afterwards, the three Lyapunov exponents can be used to evaluate the Kaplan–Yorke dimension D_{KY}, which can be obtained by evaluating (3.20), where k is an integer such that the sum of all the Lyapunov exponents (λ_i) is nonnegative. Therefore, if chaotic behavior is guaranteed in (3.1)–(3.3), then $k = 2$, so that λ_{k+1} is the third Lyapunov exponent, and the dimension D_{KY} must be higher than 2.

$$D_{KY} = k + \frac{\sum_{i=1}^{k} \lambda_i}{\lambda_{k+1}} \tag{3.20}$$

For fractional-order chaotic oscillators with more than three Lyapunov exponents, all of them must be ordered from the most positive to the less negative to evaluate D_{KY}, which is conjectured to coincide with the information dimension [31].

The Lyapunov exponents of the three chaotic oscillators were obtained by applying Wolf's method [8] and TISEAN [16], and the initial conditions for

---

Algorithm 3 Maximizing LE+ applying particle swarm optimization

---

Require: D, G, N_p, α, β, and $func(\bullet)$.

 1: Initialize the position of the particles randomly (x)
 2: Initialize the velocity of the particles (v)
 3: Evaluate the position of the particles associated with the chaotic oscillator
    described as $func(x)$. This step performs the following operations:
    Obtain the equilibrium points and evaluate the Jacobian associated with the fractional-order
    chaotic oscillator to:
    Compute the eigenvalues to estimate h
    Verify the condition of the fractional derivative with (3.13) and (3.14)
    Evaluate LE+ applying TISEAN or Wolf's method [8]
 4: Save the evaluation results in $score$ and $p \leftarrow x$
 5: Find the best value from p and save it in g
 6: **for** $(counter = 1; counter \leq G; counter++)$ **do**
 7:     **for** $(i = 1; i \leq N_p; i++)$ **do**
 8:         **for** $(j = 1; j \leq D; j++)$ **do**
 9:             $v_{ij} \leftarrow \alpha v_{ij} + U(0, \beta)\left(p_{ij} - x_{ij}\right) + U(0, \beta)\left(g_j - x_{ij}\right)$
                This evaluates the new velocity using (3.18)
10:             $x_{ij} \leftarrow x_{ij} + v_{ij}$
                This evaluates the new position using (3.19)
11:         **end for**
12:         $f_x \leftarrow func(x_i)$
13:         **if** f_x is better than $score_i$ **then**
14:             $score_i \leftarrow f_x$
15:             $p_i \leftarrow x_i$
16:             **if** p_i is better than g **then**
17:                 $g \leftarrow p_i$
18:             **end if**
19:         **end if**
20:     **end for**
21: **end for**
22: **return** x, p, g, and $score = 0$

---

the three state variables $((x(0), y(0), z(0))$ matter to reduce computing time in evaluating Lyapunov exponents and D_{KY}. In this manner, the appropriate initial conditions for each fractional-order chaotic oscillator are: $(0.1, 0.1, 0.1)$ for Lorenz described by (3.1), $(-9, -5, 14)$ for Chen described by (3.2), and $(0.5, 1, 0.5)$ for Heart Shape described by (3.3).

In order to appreciate the behavior of the metaheuristics, both DE and PSO algorithms were executed with the same conditions for the three fractional-order chaotic oscillators, i.e., the same number of populations (P) and maximum generations (G). The numerical simulations were performed for 10,000 iterations, discarding the first 1000 iterations as they include the transient behavior. In this manner, the optimization was run to maximize D_{KY} with the following conditions in both DE and PSO: $G = 20$, $P = 30$, and the search space $0.001 \leq \sigma \leq 60$; $0.001 \leq \rho \leq 180$; $0.001 \leq \beta \leq 30$ for the Lorenz system; $G = 20$, $P = 30$, and the search spaces $0.001 \leq a \leq 60$; $0.001 \leq b \leq 10$; $0.001 \leq c \leq 70$ for Chen and $G = 20$, $P = 30$, and the search spaces $0.001 \leq m \leq 10$; $0.001 \leq r \leq 10$ for

Heart Shape. The optimization results provided by both metaheuristics are given in Table 3.4, where it can be appreciated that $D_{KY} > 2$ in all cases and the maximum variation is approximately ± 0.05, which demonstrates the usefulness of applying metaheuristics like DE and PSO algorithms.

In Table 3.4 one can appreciate the feasible optimized results provided by DE and PSO algorithms for the three fractional-order chaotic oscillators (3.1)–(3.3), in which the highest values of D_{KY} are given in Table 3.5, that also include the

Table 3.4 Results of ten runs performed by differential evolution (DE) and particle swarm optimization (PSO) for the three fractional-order chaotic oscillators

Oscillator	Runs	DE			PSO		
		Max D_{KY}	Mean	σ	Max D_{KY}	Mean	σ
Lorenz	1	2.0754	2.0773	0.01486	2.0719	2.0713	0.00969
	2	2.0730	2.0732	0.01481	2.0718	2.0712	0.00969
	3	2.0843	2.0783	0.01466	2.0706	2.0703	0.00969
	4	2.0791	2.0775	0.01478	2.0712	2.0705	0.00969
	5	2.0785	2.0772	0.01474	2.0712	2.0702	0.00969
	6	2.0839	2.0770	0.01486	2.0662	2.0710	0.00969
	7	2.0796	2.0752	0.01487	2.0690	2.0702	0.00969
	8	2.0739	2.0738	0.01479	2.0684	2.0708	0.00969
	9	2.0733	2.0733	0.01451	2.0692	2.0703	0.00969
	10	2.0741	2.0740	0.01431	2.0714	2.0744	0.00969
Chen	1	2.0781	2.0789	0.02080	2.0764	2.0753	0.02059
	2	2.0791	2.0788	0.02079	2.0767	2.0753	0.02036
	3	2.0781	2.0787	0.02071	2.0767	2.0759	0.02036
	4	2.0788	2.0787	0.02075	2.0716	2.0765	0.02073
	5	2.0785	2.0787	0.02063	2.0740	2.0738	0.02077
	6	2.0786	2.0785	0.02026	2.0766	2.0715	0.02039
	7	2.0781	2.0788	0.02043	2.0766	2.0734	0.02077
	8	2.0782	2.0788	0.02031	2.0754	2.0751	0.02031
	9	2.0786	2.0789	0.02051	2.0754	2.0752	0.02054
	10	2.0789	2.0787	0.02070	2.0737	2.0736	0.02045
Heart shape	1	2.0250	2.0395	0.02036	2.0235	2.0230	0.03285
	2	2.0350	2.0377	0.02043	2.0225	2.0224	0.03286
	3	2.0510	2.0364	0.02046	2.0270	2.0210	0.03286
	4	2.0300	2.0374	0.02048	2.0229	2.0240	0.03289
	5	2.0500	2.0381	0.02038	2.0198	2.0220	0.03285
	6	2.0203	2.0383	0.02039	2.0269	2.0240	0.03288
	7	2.0203	2.0363	0.02028	2.0233	2.0230	0.03288
	8	2.0204	2.0373	0.02044	2.0170	2.0240	0.03287
	9	2.0204	2.0405	0.02026	2.0254	2.0230	0.03285
	10	2.0055	2.0414	0.02039	2.0265	2.0220	0.03287

Table 3.5 Design parameters of the five highest values of D_{KY} from Table 3.4, for each metaheuristic and fractional-order chaotic oscillator, and their corresponding LE+ value

Oscillator	DE Design parameters	LE$_+$	D_{KY}	PSO Design parameters	LE$_+$	D_{KY}
Lorenz	$\sigma = 29.9226 \; \rho = 89.8095 \; \beta = 13.9727$	3.3129	2.08430	$\sigma = 30.0000 \; \rho = 90 \; \beta = 12.3872$	3.3129	2.07230
	$\sigma = 29.9388 \; \rho = 89.8923 \; \beta = 14.1895$	3.3122	2.08390	$\sigma = 29.8297 \; \rho = 90 \; \beta = 13.7954$	3.3122	2.07290
	$\sigma = 29.7786 \; \rho = 89.7268 \; \beta = 13.4876$	3.3168	2.07960	$\sigma = 29.9966 \; \rho = 90 \; \beta = 13.4876$	3.3168	2.07190
	$\sigma = 29.9222 \; \rho = 89.9781 \; \beta = 14.1956$	3.3149	2.07910	$\sigma = 30.0000 \; \rho = 90 \; \beta = 14.1956$	3.3149	2.07197
	$\sigma = 29.7066 \; \rho = 89.8899 \; \beta = 13.7180$	3.3199	2.07410	$\sigma = 29.8375 \; \rho = 90 \; \beta = 13.8036$	3.3199	2.07179
Chen	$a = 0.3609 \; b = 0.1000 \; c = 11.3470$	0.02711	2.07890	$a = 0.3609 \; b = 0.1000 \; c = 11.3470$	0.02711	2.02700
	$a = 0.3947 \; b = 0.5490 \; c = 9.12060$	0.02600	2.07140	$a = 0.3947 \; b = 0.5490 \; c = 9.12060$	0.02600	2.02690
	$a = 0.3720 \; b = 0.2055 \; c = 12.0147$	0.02710	2.07870	$a = 0.3947 \; b = 0.2055 \; c = 9.12060$	0.02600	2.02650
	$a = 0.3930 \; b = 0.8505 \; c = 13.0501$	0.02645	2.07810	$a = 0.3930 \; b = 0.8505 \; c = 13.0501$	0.02645	2.02350
	$a = 0.3643 \; b = 0.1537 \; c = 12.7643$	0.02764	2.07880	$a = 0.3643 \; b = 0.1537 \; c = 12.7643$	0.02764	2.02330
Heart shape	$m = 0.6609 \; r = 0.100$	0.4711	2.07890	$m = 0.3609 \; r = 0.1000$	0.4711	2.02700
	$m = 0.5947 \; r = 0.549$	0.4600	2.07140	$m = 0.3947 \; r = 0.5490$	0.4600	2.02690
	$m = 0.5720 \; r = 0.205$	0.4710	2.07870	$m = 0.3947 \; r = 0.2055$	0.4600	2.02650
	$m = 0.5930 \; r = 0.850$	0.4645	2.07810	$m = 0.3930 \; r = 0.8505$	0.4645	2.02350
	$m = 0.4643 \; r = 0.153$	0.4764	2.07880	$m = 0.3643 \; r = 0.1537$	0.4764	2.02330

respective design variables or coefficients of the mathematical models associated with the best five values of D_{KY} provided by DE and PSO, and it also shows their related LE+.

It is then very convenient having different possibilities of implementing a fractional-order chaotic oscillator that may generate the same behavior in characteristics like LE+ and D_{KY}. Some of those feasible solutions may be more sensitive than others for simulating the chaotic behavior for long-times. In addition, the optimization results provided by both metaheuristics, DE and PSO, depend on the numerical approximation. As shown in Chap. 1, for integer-order chaotic oscillators, the most simple numerical method Forward Euler generates high error with a time step similar to a more elaborated method like the multistep or Runge–Kutta families. This is also a problem in simulating fractional-order chaotic oscillators, because each approximation requires different conditions. For example, using Grünwald–Letnikov it will be more easy to apply the short memory principle to reduce hardware resources while computing a similar exactness of the chaotic time series data. But, this is not the case in applying Adams–Bashforth–Moulton with the short memory principle, because the binomial coefficients in Grünwald–Letnikov tend to decrease, and in a predictor-corrector approximation they have unstable behavior. This will be detailed in Chap. 5 when describing the FPGA-based implementation of several fractional-order chaotic oscillators.

References

1. I. Grigorenko, E. Grigorenko, Chaotic dynamics of the fractional Lorenz system. Phys. Rev. Lett. **91**(3), 034101 (2003)
2. J. Lü, G. Chen, A new chaotic attractor coined. Int. J. Bifurcation Chaos **12**(3), 659–661 (2002)
3. M.F. Tolba, A.M. AbdelAty, N.S. Soliman, L.A. Said, A.H. Madian, A.T. Azar, A.G. Radwan, FPGA implementation of two fractional order chaotic systems. AEU-Int. J. Electron. Commun. **78**, 162–172 (2017)
4. G. Cardano, T.R. Witmer, *Ars Magna or the Rules of Algebra*. Dover Books on Advanced Mathematics (Dover, New York, 1968)
5. R. Garrappa, Short tutorial: solving fractional differential equations by Matlab codes. Department of Mathematics, University of Bari (2014)
6. M.-F. Danca, N. Kuznetsov, Matlab code for Lyapunov exponents of fractional-order systems. Int. J. Bifurcation Chaos **28**(5), 1850067 (2018)
7. A.D. Pano-Azucena, E. Tlelo-Cuautle, G. Rodriguez-Gomez, L.G. De la Fraga, FPGA-based implementation of chaotic oscillators by applying the numerical method based on trigonometric polynomials. AIP Adv. **8**(7), 075217 (2018)
8. A. Wolf, J.B. Swift, H.L. Swinney, J.A. Vastano, Determining Lyapunov exponents from a time series. Phys. D Nonlinear Phenomena **16**(3), 285–317 (1985)
9. A. Bespalov, N. Polyakhov, Determination of the largest Lyapunov exponents based on time series. World Appl. Sci. J. **26**(2), 157–164 (2013)
10. K. Diethelm, *The Analysis of Fractional Differential Equations: An Application-Oriented Exposition Using Differential Operators of Caputo Type* (Springer, New York, 2010)
11. K. Diethelm, N.J. Ford, A.D. Freed, A predictor-corrector approach for the numerical solution of fractional differential equations. Nonlinear Dynam. **29**(1–4), 3–22 (2002)

12. R. Garrappa, Predictor-corrector PECE method for fractional differential equations. MATLAB Central File Exchange [File ID: 32918] (2011)
13. M.-F. Danca, N. Kuznetsov, Matlab code for Lyapunov exponents of fractional-order systems. Int. J. Bifurcation Chaos **28**(5), 1850067 (2018)
14. H. Kantz, A robust method to estimate the maximal Lyapunov exponent of a time series. Phys. Lett. A **185**(1), 77–87 (1994)
15. M.T. Rosenstein, J.J. Collins, C.J. De Luca, A practical method for calculating largest Lyapunov exponents from small data sets. Phys. D Nonlinear Phenomena **65**(1–2), 117–134 (1993)
16. R. Hegger, H. Kantz, T. Schreiber, Practical implementation of nonlinear time series methods: the TISEAN package. Chaos: An Interdisciplinary J. Nonlinear Sci. **9**(2), 413–435 (1999)
17. J. Petržela, Optimal piecewise-linear approximation of the quadratic chaotic dynamics. Radioengineering **21**(1), 20–28 (2012)
18. S.-Y. Li, S.-C. Huang, C.-H. Yang, Z.-M. Ge, Generating tri-chaos attractors with three positive Lyapunov exponents in new four order system via linear coupling. Nonlinear Dynam. **69**(3), 805–816 (2012)
19. Y. Sun, C.Q. Wu, A radial-basis-function network-based method of estimating Lyapunov exponents from a scalar time series for analyzing nonlinear systems stability. Nonlinear Dynam. **70**(2), 1689–1708 (2012)
20. C. Li, J.C. Sprott, Amplitude control approach for chaotic signals. Nonlinear Dynam. **73**(3), 1335–1341 (2013)
21. K.S. Miller, B. Ross, *An Introduction to the Fractional Calculus and Fractional Differential Equations* (Wiley, Hoboken, 1993)
22. K. Oldham, J. Spanier, *The Fractional Calculus Theory and Applications of Differentiation and Integration to Arbitrary Order*, vol. 111 (Elsevier, Amsterdam, 1974)
23. M. Caputo, Linear models of dissipation whose q is almost frequency independent-II. Geophys. J. Int. **13**(5), 529–539 (1967)
24. L. Dorcak, J. Prokop, I. Kostial, Investigation of the properties of fractional-order dynamical systems, in *Proceedings of 11th International Conference on Process Control* (1994), pp. 19–20
25. D. Cafagna, G. Grassi, On the simplest fractional-order memristor-based chaotic system. Nonlinear Dynam. **70**(2), 1185–1197 (2012)
26. R. Storn, K. Price, Differential evolution – a simple and efficient heuristic for global optimization over continuous spaces. J. Global Optim. **11**(4), 341–359 (1997)
27. R. Kennedy, J. Eberhart, Particle swarm optimization, in *Proceedings of IEEE International Conference on Neural Networks IV*, vol. 1000 (1995)
28. C.J. Yang, W.D. Zhu, G.X. Ren, Approximate and efficient calculation of dominant Lyapunov exponents of high-dimensional nonlinear dynamic systems. Commun. Nonlinear Sci. Numer. Simul. **18**(12), 3271–3277 (2013)
29. L. Dieci, Jacobian free computation of Lyapunov exponents. J. Dyn. Diff. Equ. **14**(3), 697–717 (2002)
30. S. Rugonyi, K.-J. Bathe, An evaluation of the Lyapunov characteristic exponent of chaotic continuous systems. Int. J. Numer. Methods Eng. **56**(1), 145–163 (2003)
31. J.L. Kaplan, J.A. Yorke, Chaotic behavior of multidimensional difference equations, in *Functional Differential Equations and Approximation of Fixed Points* (Springer, Berlin, 1979), pp. 204–227

Chapter 4
Analog Implementations of Fractional-Order Chaotic Systems

4.1 Implementation of Fractional-Order Chaotic Oscillators Using Amplifiers and Fractors

In the recent literature, researchers have introduced approximations in the frequency and time domains for the simulation of fractional-order systems. In the case of fractional-order chaotic oscillators, preliminar works can be revised in [1, 2]. Time domain methods were described in Chap. 1, and will be applied in Chap. 5 for the FPGA-based implementation of several fractional-order chaotic oscillators. This chapter discusses frequency domain methods to approximate the fractional operator as shown in [3], and they will be applied to approximate the solution of fractional-order chaotic oscillators, as shown in [4].

Fractional-order systems can be considered as a generalization of integer-order ones, so that the most common instance of a fractional-order system can be represented by a transfer function in the frequency domain. It can be done considering the irrational transfer function given in (4.1), which is called as fractional-order integrator and it can be found in many physical phenomena. In such approximation of the fractional-order integrator, $s = j\omega$ becomes to be the complex frequency and is a positive real number such that $0 < \alpha < 1$.

$$H(s) = \frac{1}{s^\alpha} \qquad \alpha \in \mathbb{R}^+ \tag{4.1}$$

The fractional-order integrator can be approached by considering the basic formulation of Charef's approximation method, which is detailed in [5]. Basically, a dynamical system can be modeled in the frequency domain by the transfer function of a single fractional power pole as given in (4.2), where pT is the pole of the fractional-order system and α is the fractional-order of the system.

© Springer Nature Switzerland AG 2020
E. Tlelo-Cuautle et al., *Analog/Digital Implementation of Fractional Order Chaotic Circuits and Applications*, https://doi.org/10.1007/978-3-030-31250-3_4

$$H(s) = \frac{1}{(1 + s/p_T)^\alpha} \tag{4.2}$$

Plotting (4.1) will provide a figure showing an estimated slope with -20α dB/dec, and with a number of zigzag straight lines connected with individual slopes of 0 dB/dec and -20α dB/dec. That way, (4.2) can be rationalized by the recursive formula described in (4.3), which can be associated with the frequency response in the range ω_{max}, at which a corner frequency p_T and the error y are given in dB between the actual and the approximated line. Moreover, the zeros and poles of the transfer function can be recursively calculated by (4.4). The first approximation of the pole and zero is given by (4.5).

$$H(s) = \frac{\Pi_{i=0}^{N-1} \left(1 + \frac{s}{z_i}\right)}{\Pi_{i=0}^{N} \left(1 + \frac{s}{p_i}\right)}, \tag{4.3}$$

$$z_{N-1} = p_{N-1} 10^{[y/10(1-\alpha)]}$$

$$p_N = z_{N-1} 10^{[y/10\alpha]}. \tag{4.4}$$

$$p_0 = p_T 10^{[y/20\alpha]}$$

$$z_0 = p_0 10^{[y/10(1-\alpha)]} \tag{4.5}$$

The next step consists on determining the value of N, and it can be done trying several values so that a specified accuracy of the approximated rational transfer function at the corner frequency can be obtained. The frequency corner p_T can be determined as it is done for the integer-order systems, i.e., at the corner measured at -3α dB. The pole p_0 can be determined by taking into account a specified error, and p_N can be determined evaluating N, where a, b are given by (4.6), in which the location ratio of a zero to a previous pole is equal to the ratio of a pole to a previous zero, respectively, and it is equal to (4.7).

$$a = 10^{[y/10(1-\alpha)]} = \frac{z_{N-1}}{p_{N-1}}$$

$$b = 10^{[y/10\alpha)]} = \frac{p_{N-1}}{z_{N-1}} \tag{4.6}$$

$$ab = 10^{y/10\alpha(1-m)} = \frac{z_{N-1}}{z_{N-2}} = \frac{p_N}{p_{N-1}}. \tag{4.7}$$

Now, N can be evaluated by the expression given in (4.8).

$$N = \text{Integer} \left[\frac{log\left(\frac{\omega_{max}}{p_0}\right)}{log(ab)} \right] + 1. \tag{4.8}$$

Several implementations of fractional-order chaotic oscillators have been introduced, for example: the authors in [6, 7] describe implementations of fractional-order systems with different orders of the derivatives, and also fractional-order hyperchaotic oscillators have been implemented in [8, 9]. For instance, Lorenz chaotic oscillator is a three-dimensional dynamical system that can be adapted to become fractional-order, as described in [10], where the integer-order system of equations given in (1.1) are adapted as shown in (4.9), where it can be appreciated that just the derivatives have fractional-orders q_1, q_2, q_3, while the rest of the equations have the same descriptions as for the integer-order model.

$$
\begin{aligned}
{}_0D_t^{q_1}x(t) &= \sigma(y(t) - x(t)), \\
{}_0D_t^{q_2}y(t) &= x(t)(\rho - z(t)) - y(t), \\
{}_0D_t^{q_3}z(t) &= x(t)y(t) - \beta z(t)
\end{aligned}
\tag{4.9}
$$

The fractional-order Lorenz chaotic oscillator described by (4.9) can be designed by setting: $\sigma = 10, \rho = 28$, and $\beta = 8/3$, but in this fractional-order case, the derivatives can have the same orders, such as $q_1 = q_2 = q_3 = 0.9$. In this case the fractional-order dynamical system is of commensurate type; otherwise, if the fractional orders are different the dynamical system is incommensurate. A time simulation of the fractional-order Lorenz system applying Grünwald–Letnikov method generates the phase-space portraits shown in Fig. 4.1. These figures show ranges of the state variables x, y, z higher than that supported by commercially available operational amplifiers; therefore, they will be down-scaled in the following sections to allow electronic circuit implementation.

Doing the simulation of the same fractional-order system given in (4.9), and considering commensurate fractional-order, then Lorenz chaotic oscillator can be solved in the frequency domain using the fractional integrator that is described by (4.1) or equivalently by $H(s) = \dfrac{1}{s^q}$, as detailed in [11].

The circuit implementation of the fractional-order Lorenz chaotic oscillator given in (4.9) requires an approximation technique for the fractional-order integrator, because the standard definition for the integer-order integrator cannot be applied and then a fractional integrator does not allow a direct circuit implementation. That way, a circuit having properties of fractional-order is called fractance or fractor [12], and it can be implemented through arrays of passive circuit elements such as resistances and capacitances that basically are interconnected in series or parallel connections [13], as shown in Fig. 4.2. The symbol of the fractance has the form of an element similar to a resistance and a capacitance.

Two types of circuits related to the implementation of the fractance using RC arrays can be observed, namely ladder fractance topology and tree fractance. In electronics, yet one cannot find a fractance device that accomplishes the desired fractional-order being implemented. Henceforth, the fractance or fractor is emulated using ladder, cascade, or tree topologies, or combinations among them. In addition, using an operational amplifier, it is possible to perform the differentiation

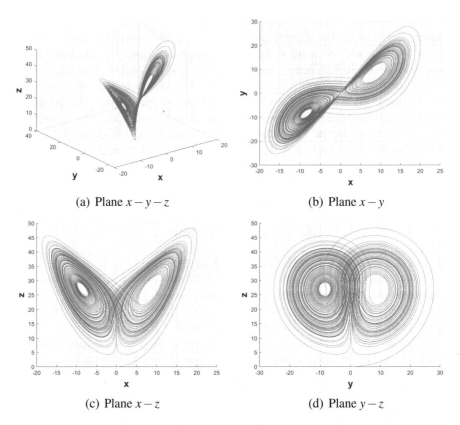

(a) Plane $x - y - z$ (b) Plane $x - y$

(c) Plane $x - z$ (d) Plane $y - z$

Fig. 4.1 Numerical simulation of the fractional-order Lorenz chaotic oscillator applying the time domain Grünwald–Letnikov method for $T_{sim} = 100$ s, a time-step of $h = 5 \times 10^{-3}$, initial conditions of $(x_0, y_0, z_0) = (0.1, 0.1, 0.1)$, and the fractional-orders of the derivatives $q_1 = q_2 = q_3 = 0.9$. (**a**) Plane $x - y - z$. (**b**) Plane $x - y$. (**c**) Plane $x - z$. (**d**) Plane $y - z$

and integration operations, as shown in Fig. 4.3. Afterwards, the fractional-order integrator and differentiator can be used in the same way as it is done in the integer-order dynamical systems.

In the Laplace domain, the circuit model between nodes A and B in Fig. 4.2a, c can be used to implement approximations of $\dfrac{1}{s^q}$ in the range of the fractional order $0.1 < q < 0.9$, at steps of 0.1, as detailed in Table 4.1, which is obtained according to [11, 14].

As one can infer, the main objective is finding the approximation of the fractional-order of the fractance [14]. The transfer function in the Laplace domain that better approaches a given fractional-order should be implemented with standard values of the capacitors, while the resistors can be tuned to generate the lowest error to emulate the desired fractance. From the approached transfer functions, the values of the R and C elements can be determined according to [2–4]. This process may

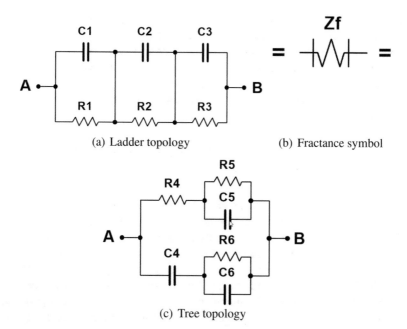

(a) Ladder topology   (b) Fractance symbol

(c) Tree topology

Fig. 4.2 RC-circuit approximation of the fundamental fractance to implement the fractional-order of a dynamical system in the range $0 < q < 1$. (**a**) Ladder topology. (**b**) Fractance symbol. (**c**) Tree topology

Fig. 4.3 Basic fractional-order differentiator and integrator implemented with an operational amplifier and a fractance emulated with RC topologies. (**a**) Basic fractional-order integrator. (**b**) Basic fractional-order differentiator

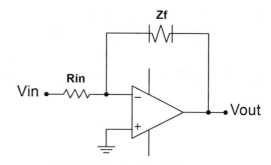

(a) Basic fractional-order integrator

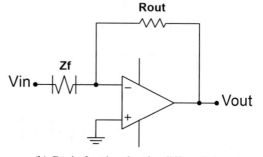

(b) Basic fractional-order differentiator

give rise to some difficulties to find the values that generate low sensitivities and tolerances, so that one must take care of the available combinations of values that are assigned to the passive circuit elements. In the following sections is shown the implementation of the transfer functions that approach a fractional-order value using FPAAs [15, 16]. This is helpful to avoid mismatches among the different amplifiers required to complete the whole circuit implementation of a chaotic oscillator.

4.2 Ladder Fractance to Implement the Fractional-Order Lorenz Chaotic Circuit

Let us consider the analog implementation of the fractional-order Lorenz chaotic oscillator with a whole fractional-order of 2.7. Therefore, as a commensurate dynamical system with $q = 0.9$, one can choose the approximation from the Laplace expressions given in Table 4.1. In this case, one can use the transfer function described by (4.10), which has a maximum discrepancy of $y = 2\,dB$ [11, 13], with respect to the exact solution of the fractional-order integrator. Using the ladder topology for interconnecting RC elements, as shown in Fig. 4.2a, and performing an analysis considering $q = 0.9$ one must find the values of the circuit elements.

$$H(s) = \frac{1}{s^{0.9}} = \frac{2.2675(s + 1.292)(s + 215.4)}{(s + 0.01292)(s + 2.154)(s + 359.4)} \tag{4.10}$$

The fractance of ladder type has three pairs of series-connected resistor-capacitor arrays. Applying Laplace operations one can obtain the transfer function $H(s)$ between the nodes A and B from Fig. 4.2a, which is given in (4.11), where C_0 is a unitary parameter, thus if one considers $C_0 = 1\,\mu F$, then $F(s) = H(s)C_0 = 1/s^{0.9}$. Afterwards, the expression of the transfer function for the ladder topology of the fractance is given by (4.12), where C_a is equivalent to (4.13).

$$H(s) = \frac{\frac{C_0}{C_1}}{s + \frac{1}{R_1 C_1}} + \frac{\frac{C_0}{C_1}}{s + \frac{1}{R_2 C_2}} + \frac{\frac{C_0}{C_1}}{s + \frac{1}{R_3 C_3}} \tag{4.11}$$

$$H(s) = \frac{1}{C_0} \frac{C_a \left[s^2 + \frac{s\left(\frac{C_2 + C_3}{R_1} + \frac{C_2 + C_3}{R2} + \frac{C_1 + C_2}{R_3}\right) + \frac{R_1 + R_2 + R_3}{R_1 R_2 R_3}}{C_1 C_2 + C_2 C_3 + C_1 C_3} \right]}{\left(s + \frac{1}{R_1 C1}\right)\left(s + \frac{1}{R_2 C_2}\right)\left(s + \frac{1}{R_3 C_3}\right)} \tag{4.12}$$

$$C_a = \left(\frac{C_0}{C_1} + \frac{C_0}{C_2} + \frac{C_0}{C_3}\right) \tag{4.13}$$

Table 4.1 Integer-order transfer functions that approximate fractional-order integrators between $0.1 < q < 0.9$, with a maximum discrepancy $y = 2\,\mathrm{dB}$ and bandwidth of the system of $\omega_{max} = 10^3$ rad/s, taken from [11]

q-Order		$H(s)$ transfer function
0.1	$\approx$	$\dfrac{1584.8932(s+0.1668)(s+27.83)}{(s+0.1)(s+16.68)(s+2783)}$
0.2	$\approx$	$\dfrac{79.4328(s+0.05623)(s+1)(s+17.78)}{(s+0.03162)(s+0.5623)(s+10)(s+177.8)}$
0.3	$\approx$	$\dfrac{39.8107(s+0.0416)(s+0.3728)(s+3.34)(s+29.94)}{(s+0.02154)(s+01931)(s+1.73)(s+15.51)(s+138.9)}$
0.4	$\approx$	$\dfrac{35.4813(s+0.03831)(s+0.26.1)(s+1.778.)(s+12.12)(s+82.54)}{(s+0.01778)(s+0.1212)(s+0.8254)(s+5.623)(s+38.31)(s+261)}$
0.5	$\approx$	$\dfrac{15.8489(s+0.03.981)(s+0.2512)(s+1.585)(s+10)(s+63.1)}{(s+0.01585)(s+01)(s+0.631)(s+3981)(s+25.12)(s+158.5)}$
0.6	$\approx$	$\dfrac{10.7978(s+0.04642)(s+0.3162)(s+2.154)(s+14.68)(s+100.)}{(s+0.01468)(s+0.1)(s+0.6813)(s+4642)(s+3162)(s+2154)}$
0.7	$\approx$	$\dfrac{9.3633(s+0.06449)(s+0.578)(s+5.179)(s+46.42)(s+416)}{(s+0.01389)(s+0.1245)(s+1.116)(s+10)(s+89.62)(s+803.1)}$
0.8	$\approx$	$\dfrac{5.3088(s+0.1334)(s+2.371)(s+42.17)(s+749.9)}{(s+0.01334)(s+0.2371)(s+4.217)(s+74.99)(s+1334)}$
0.9	$\approx$	$\dfrac{2.2675(s+1.292)(s+215.4)}{(s+0.01292)(s+2.154)(s+359.4)}$

Table 4.2 Proposed values of the passive circuit elements for the circuit implementation of the fractional-order Lorenz chaotic oscillator

Circuit element	Element value	Circuit element	Element value
$R_1 = R_{11} = R_{19}$	62.84 MΩ	$R_4 = R_{10} = R_{15}$	
		$= R_{18} = R_{23}$	1 kΩ
$R_2 = R_{12} = R_{20}$	250 kΩ	$R_3 = R_{13} = R_{21}$	2.5 kΩ
$R_5 = R_8 = R_9 = R_{16}$			
$= R_{24} = R_{25} = R_{26}$	10 kΩ	$R_6 = R_7$	2 kΩ
R_{17}	715 Ω	R_{14}	20 kΩ
R_{22}	7.5 kΩ	$C_1 = C_4 = C_7$	1.23 μF
$C_2 = C_5 = C_8$	1.84 μF	$C_3 = C_6 = C_9$	1.1 μF

Comparing (4.10) and (4.12) is helpful to obtain the values of the resistors and capacitors of the fractance required to implement (4.9). The values that are proposed for this implementation are listed in Table 4.2, which includes the values of other RC elements to implement the whole fractional-order Lorenz chaotic oscillator that is detailed by the figures shown below.

According to the circuit model shown in Fig. 4.2a, it can be used to implement the circuit of the 2.7-order Lorenz chaotic oscillator. Looking to the time simulation applying Grünwald–Letnikov method, one can see that the phase-space portraits shown in Fig. 4.1 exceed the range of values supported by commercially available operational amplifiers, as well as of the required analog multipliers. In this manner, one must down-scale those state variable ranges to implement (4.14).

$$\frac{d^{0.9}x}{dt^{0.9}} = \sigma(y - x),$$

$$\frac{d^{0.9}y}{dt^{0.9}} = x(\rho - z) - y, \tag{4.14}$$

$$\frac{d^{0.9}z}{dt^{0.9}} = xy - \beta z$$

Applying circuit theory and the guidelines for down-scaling given in [17], then the updated equations are given in (4.15). These equations can be implemented using the RC-approached fractance, the amplifiers like AD712 / LM324 / TL082, and the multiplier AD633, which needs an output coefficient value of 0.1 and a power supply of ± 12 V.

$$\frac{X(s)}{H(s)} = \frac{R_5}{C_o R_4} \left[\frac{Y(s)}{R_7} - \frac{R_8 X(s)}{R_9 R_4} \right]$$

$$\frac{Y(s)}{H(s)} = \frac{R_{13}}{C_o R_{14}} \left[\frac{X(s)}{R_{10}} - \frac{k R_{28} L\{x(t)z(t)\}}{R_{27} R_{12}} - \frac{Y(s)}{R_7} \right] \tag{4.15}$$

$$\frac{Z(s)}{H(s)} = \frac{R_{22}}{C_o R_{23}} \left[\frac{k L\{x(t)y(t)\}}{R_{21}} - \frac{R_{25} Z(s)}{R_{27} R_{20}} \right]$$

In (4.15), k represents the output coefficient value of the multiplier AD633, and $H(s) = \frac{1}{s^{0.9}}$. The following figures detail the analog implementation of the whole fractional-order Lorenz chaotic oscillator. For instance, Fig. 4.4 details the implementation of state variable x, Fig. 4.5 details the implementation of state variable y, and Fig. 4.6 details the building blocks implementing state variable z, accomplishing fractional-order equal to 2.7, thus all the fractional derivatives have the same fractional-order $q = 0.9$.

Figure 4.7 shows the phase-space portraits of the simulation of the whole circuit of the 2.7-order Lorenz chaotic oscillator. It can be appreciated that the numerical simulation using electronic devices and RC-ladder topologies to implement the fractance are in good agreement with the simulation of the same 2.7-order Lorenz chaotic oscillator applying the time domain Grünwald–Letnikov method, and shown in Fig. 4.1. Thus demonstrating that the approximation of the fractional-order integrator in the frequency domain is also quite useful, and provides similar simulation results when applying time domain methods. In addition, the implementation of the

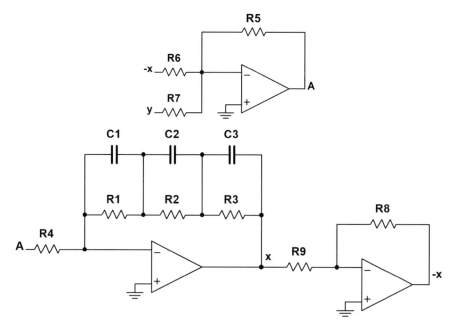

Fig. 4.4 OpAmp-based topology implementing the state variable x

resulting Laplace expressions is more easy to design using electronic devices than the time domain methods, which require more digital hardware resources, as shown in Chap. 5.

4.3   Tree Fractance to Implement the Fractional-Order Lorenz Chaotic Circuit

Similar to the frequency domain approximation of the fractance by using RC-elements in ladder/cascade connection, this section shows the application of the tree topology to implement the fractance for the 2.7-order Lorenz chaotic oscillator, i.e., setting $q_1 = q_2 = q_3 = 0.9$ in (4.9). The design of the electronic circuit blocks is performed in a similar way as above, as well as the simulation is performed using a circuit simulator like multisim. The authors in [1] presented a design of this fractional-order chaotic oscillator, and herein details on the implementation using amplifiers is given.

Let us choose the transfer function that approaches the fractional order of the integrator with $q = 0.9$ from Table 4.1, which $H(s)$ is given in (4.10), and then it can be implemented now using the tree topology of RC elements shown in Fig. 4.2c. Therefore, applying circuit theory in the Laplace domain, the transfer function $H(s)$ between the nodes A and B in Fig. 4.2c is given by (4.16), which can be expanded

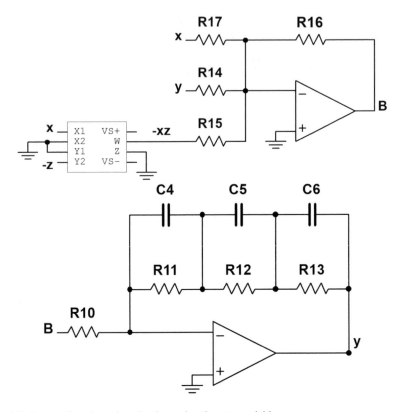

Fig. 4.5 OpAmp-based topology implementing the state variable y

to get (4.17), where C_s is given by (4.18). The transfer function given in (4.17) is different to the approximation using a ladder fractance, and where C_0 again is a unitary parameter, thus $C_0 = 1\mu F$ and $F(s) = H(s)C_0 = 1/s^{0.9}$. Comparing (4.10) and (4.17), one can find the values of the RC elements. The tree fractance is then implemented using the values listed in Table 4.3, for the R and C circuit elements connected in a tree topology.

$$H(s) = \left[R_1 + \left(R_2 // \frac{1}{sC_2} \right) \right] // \left[\frac{1}{sC_1} + \left(R_3 // \frac{1}{sC_3} \right) \right] \tag{4.16}$$

$$H(s) = \frac{1}{C_0} \frac{\left(\dfrac{C_0}{C_1} + \dfrac{C_0}{C_3} \right) \left(s + \dfrac{R_1 + R_2}{R_1 C_2 R_2} \right) \left(s + \dfrac{1}{C_1 R_3 + C_3 R3} \right)}{s^3 + s^2 \left(\dfrac{R_1 + R_2}{R_1 C_2 R_2} + \dfrac{1}{C_3 R_3} + \dfrac{C_1 + C_3}{C_1 R_1 C_3} \right) + sC_s + \dfrac{1}{C_1 R_1 C_2 R_2 C_3 R_3}} \tag{4.17}$$

$$C_s = \left(\frac{R_1 + R_2}{R_1 C_2 R_2 C_3 R_3} + \frac{1}{C_1 R_1 C_3 R_3} + \frac{C_1 + C_3}{C_1 R_1 C_2 R_2 C_3} \right) \tag{4.18}$$

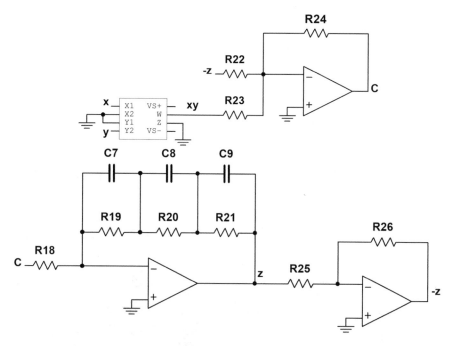

Fig. 4.6 OpAmp-based topology implementing the state variable z

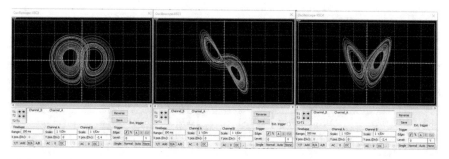

Fig. 4.7 Phase-space portraits of the 2.7-order chaotic attractor when implementing the Lorenz system as a commensurate oscillator with $q_1 = q_2 = q_3 = 0.9$, and observing the plots with 1 V/Div

Doing the same process to find the values of the circuit elements R and C to approximate $H(s) = \dfrac{1}{s^q}$ with fractional-order values between $0.1 < q < 0.9$, at steps of 0.1, as shown in [14], one can use the values listed in Tables 4.4 and 4.5.

The tree topology is helpful to implement an electronic circuit of a fractional-order chaotic oscillator, as the Lorenz one having fractional-order equal to 2.7. As the ranges of the state variables exceed the values that can provide an amplifier and multiplier, one must down-scale the mathematical model given in (4.9), the updated

Table 4.3 Values of the RC elements to implement the tree fractance and the whole 2.7-order Lorenz chaotic oscillator

Circuit element	Element value	Circuit element	Element value
$R_1 = R_{11} = R_{19}$	1.55 MΩ	$R_4 = R_{10} = R_{15}$ $= R_{18} = R_{23}$	1 kΩ
$R_2 = R_{12} = R_{20}$	62.84 MΩ	$R_3 = R_6 = R_7$ $= R_{13} = R_{21}$	2.5 kΩ
$R_5 = R_8 = R_9 = R_{16}$ $= R_{17} = R_{24} = R_{25} = R_{26}$	10 kΩ	R_{14}	7 kΩ
R_{22}	33.5 kΩ	$C_3 = C_6 = C_9$	1.1 μF
$C_1 = C_4 = C_7$	0.73 μF	$C_2 = C_5 = C_8$	0.52 μF

Table 4.4 Approximated values of Rs to implement the tree fractance of $1/s^q$ (2 dB)

q	R_1 (MΩ)	R_2 (MΩ)	R_3 (MΩ)	R_4 (MΩ)	R_5 (MΩ)	R_6 (MΩ)
0.1	0.9517	0.6332	1.363			
0.2	0.5454	0.7553	3.307	1.211		
0.3	0.3476	1.296	4.854	2.339	0.2196	
0.4	0.1845	0.4294	2.080	5.698	0.7819	1.961
0.5	0.1749	0.4417	2.480	9.410	0.6957	2.354
0.6	0.08291	0.3202	2.594	15.45	0.5436	2.853
0.7	0.02169	0.1205	2.016	24.99	0.2806	3.247
0.8	0.008586	0.09269	6.325	39.69	0.4900	
0.9	1.550	61.54	0.002526			

Table 4.5 Approximated values of Cs to implement the tree fractance of $1/s^q$ (2 dB)

q	C_1 (μF)	C_2 (μF)	C_3 (μF)	C_4 (μF)	C_5 (μF)	C_6 (μF)
0.1	0.02572	15.77	0.0006468			
0.2	5.366	0.1814	0.01262	1.560		
0.3	4.927	0.5417	0.02525	2.752	0.3098	
0.4	5.023	0.4836	0.02834	0.1456	0.1428	4.854
0.5	3.793	0.5827	0.06416	0.1751	0.2292	4.441
0.6	2.741	0.5489	0.09585	0.1692	0.2442	3.247
0.7	1.917	0.4657	0.1131	0.1500	0.1964	2.019
0.8	0.9503	0.6139	0.2349	0.2337	0.2391	
0.9	0.7346	0.5221	1.1030			

circuital equations are given in (4.15), where again k is the output coefficient of the multiplier AD633 to implement $H(s) = \dfrac{1}{s^{0.9}}$. Figure 4.8 shows the implementation of the state variable x and using the tree topology to implement the fractance, Fig. 4.9 shows the implementation of state variable y, and Fig. 4.10 shows the implementation for state variable z. The simulation results to observe the phase-space portraits of the fractional-order Lorenz chaotic oscillator are shown in Fig. 4.11.

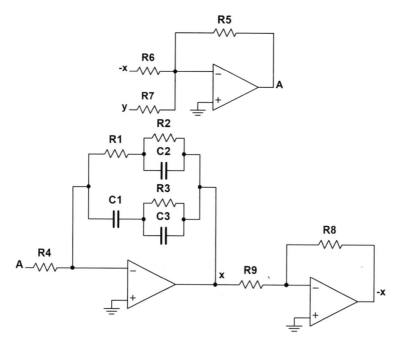

Fig. 4.8 Implementation of the 2.7-order Lorenz chaotic circuit for variable x, using tree fractance

As one can observe in Fig. 4.11, the simulation results are quite similar as those ones applying the Grünwald–Letnikov method and using the ladder topology to implement the fractance. This confirms that one can approach the fractance using ladder or tree topologies of RC elements arrays, or one can explore the implementation using other topology or a combination of both.

4.4   Implementation of Fractional-Order Chaotic Oscillators Using FPAAs

Implementing fractional-order chaotic oscillators using amplifiers, multipliers, and RC elements can be tedious and the topologies may have high sensitivities. One way to diminish mismatches among the amplifiers and RC elements can be by using FPAAs, which include amplifiers and passive R and C elements. For example, the authors in [15, 16, 18] introduced implementations of the fractional-order integrators using FPAAs. Those cases are based on other approximations, different to the one detailed above.

Let us consider the fractional-order integrator of order $q = 0.9$, which is implemented using the ladder RC topology to have the topology shown in Fig. 4.12.

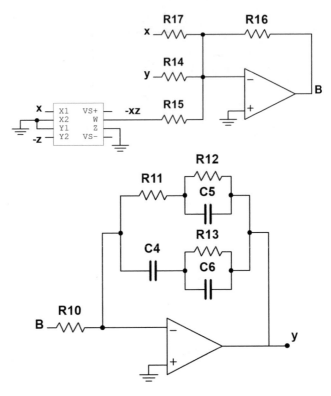

Fig. 4.9 Implementation of the 2.7-order Lorenz chaotic circuit for variable y, using tree fractance

Its transfer function can be generated performing circuit analysis to get (4.19), and which can be expanded to obtain (4.20), which has a second Laplace order in its numerator and a third order in its denominator. These expressions can be implemented using three configurable analog modules (CAMs) into an FPAA.

$$T(s) = \frac{V_o(s)}{V_i(s)} = \frac{Z_o}{Z_i} = \frac{\left(\dfrac{R_1}{1+sR_1C_1}\right) + \left(\dfrac{R_2}{1+sR_2C_2}\right) + \left(\dfrac{R_3}{1+sR_3C_3}\right)}{R_{in}}$$

$$(4.19)$$

$$T(s) = \frac{1}{R_{in}} \left[\frac{\begin{array}{c} R_1(1+sR_2C_2)(1+sR_3C_3) + R_2(1+sR_1C_1)(1+sR_3C_3) \\ + R_3(1+sR_1C_1)(1+sR_2C_2) \end{array}}{(1+sR_1C_1)(1+sR_2C_2)(1+sR_3C_3)} \right]$$

$$(4.20)$$

The FPAA-based implementation of the expression in (4.20) can be described in Anadigm Designer 2 electronic design automation (EDA) software, which is free

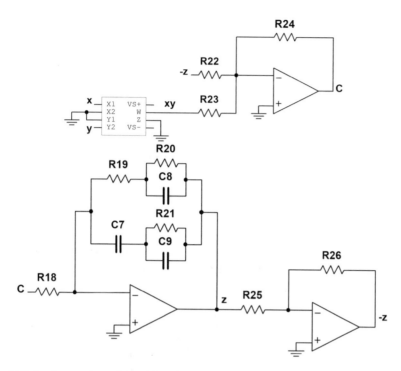

Fig. 4.10 Implementation of the 2.7-order Lorenz chaotic circuit for variable z, using tree fractance

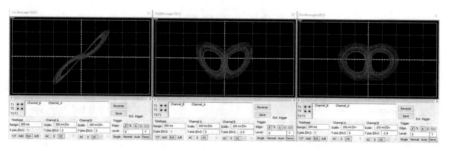

Fig. 4.11 Simulation results of the 2.7-order Lorenz chaotic oscillator when $q_1 = q_2 = q_3 = 0.9$, and observed with 1 V/Div

and available online. One should review the options of the FPAA for implementing biquadratic/bilinear filters and also single pole/zero topologies to find the best choice for synthesizing the desired Laplace expression. In this case, one is interested to approach the transfer function obtained from the fractional-order integrator shown in Fig. 4.12. That way, beginning from (4.20), one can see that one can choose the implementation of the single pole/zero thus requiring three CAMs. They are sketched in (4.21) to implement a biquadratic block, in (4.22) modeling a single

Fig. 4.12 Fractional-order
integrator based on ladder
fractance

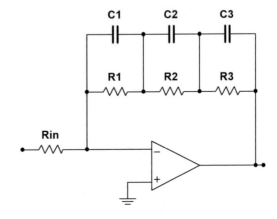

pole/zero bilinear filter, and in (4.23) to describe a lowpass bilinear filter. One can
perform a cascade connection to synthesize (4.20), using Anadigm Designer 2 EDA
tool [19].

$$T(s) = \frac{G_H \left(s^2 + \dfrac{w_z}{Q_z} s + w_z^2 \right)}{\left(s^2 + \dfrac{w_p}{Q_p} s + w_p^2 \right)} \tag{4.21}$$

$$T(s) = \frac{G_H (s + w_z)}{(s + w_p)} \tag{4.22}$$

$$T(s) = \frac{w_o G}{(s + w_o)} \tag{4.23}$$

Expanding the transfer function given in (4.20) one gets (4.24), which can be
also expressed by (4.25). In this expression one can identify the natural frequencies
of the transfer function, and they are assigned to: $\tau_1 = R_1 C_1$, $\tau_2 = R_2 C_2$, and
$\tau_3 = R_3 C_3$, which can be substituted in (4.25) to obtain the most familiar expression
given in (4.26).

$$T(s) = \frac{1}{R_{in}} \left[\frac{\begin{array}{c} R_1 \left[1 + s(R_2 C_2 + R_3 C_3) + s^2 (R_2 C_2 R_3 C_3) \right] \\ + R_2 \left[1 + s(R_1 C_1 + R_3 C_3) + s^2 R_1 C_1 R_3 C_3 \right] \\ + R_3 \left[1 + s(R_1 C_1 + R_2 C_2) + s^2 R_1 C_1 R_2 C_2 \right] \end{array}}{R_1 C_1 R_2 C_2 R_3 C_3 \left(\dfrac{1}{R_1 C_1} + s \right) \left(\dfrac{1}{R_2 C_2} + s \right) \left(\dfrac{1}{R_3 C_3} + s \right)} \right] \tag{4.24}$$

$$T(s) = \frac{1}{R_{in} R_1 C_1 R_2 C_2 R_3 C_3}$$

$$\times \left[\frac{\begin{array}{c} R_1 + R_2 + R_3 + s[R_1(R_2 C_2 + R_3 C_3) \\ + R_2(R_1 C_1 + R_3 C_3) + R_3(R_1 C_1 + R_2 C_2)] \\ + s^2 R_1 R_2 R_3 (C_2 C_3 + C_1 C_3 + C_1 C_2)] \end{array}}{\left(\dfrac{1}{R_1 C_1} + s \right) \left(\dfrac{1}{R_2 C_2} + s \right) \left(\dfrac{1}{R_3 C_3} + s \right)} \right] \qquad (4.25)$$

$$T(s) = \frac{1}{R_{in} \tau_1 \tau_2 \tau_3} \left[\frac{\begin{array}{c} R_1 + R_2 + R_3 \\ + s[R_1(\tau_2 + \tau_3) + R_2(\tau_1 + \tau_3) + R_3(\tau_1 + \tau_2)] \\ + s^2(R_1 \tau_2 \tau_3 + R_2 \tau_1 \tau_3 + R_3 \tau_1 \tau_2) \end{array}}{\left(\dfrac{1}{\tau_1} + s \right) \left(\dfrac{1}{\tau_2} + s \right) \left(\dfrac{1}{\tau_3} + s \right)} \right]$$

$$(4.26)$$

The symbolic expression given in (4.26) includes three time constants τ_1, τ_2, τ_3 that can be synthesized into an FPAA just by using two pole/zero bilinear filters and one lowpass bilinear filter. They can be connected in cascade to approach the transfer function described in (4.27), where $w = 2\pi f$, so that the expressions given in (4.28) arise. The values related to (4.28) determine the pole and zero frequencies being synthesized by the CAMs, so that one can express the equivalent transfer function given in (4.29). Also, reorganizing (4.26) one gets (4.30). Comparing the symbolic expressions given in (4.29) and (4.30), one can associate the approached values for the poles of the bilinear filters, which are given in (4.31), where f_{p1} is associated with the pole frequency of the first bilinear filter, f_{p2} is the pole frequency of the second bilinear filter, and $w_o = 2\pi f_o$, where f_o is the cut-off frequency of the lowpass filter.

$$T(s) = \frac{G_{H1}(w_{z1} + s)}{(w_{p1} + s)} \cdot \frac{G_{H2}(w_{z2} + s)}{(w_{p2} + s)} \cdot \frac{w_o G}{w_o + s} \qquad (4.27)$$

$$w_{p1} = \frac{1}{\tau_1}, \qquad w_{p2} = \frac{1}{\tau_2}, \qquad w_o = \frac{1}{\tau_3} \qquad (4.28)$$

$$T(s) = w_o G_{H1} G_{H2} G \left[\frac{w_{z1} w_{z2} + s(w_{z1} + w_{z2}) + s^2}{(w_{p1} + s)(w_{p2} + s)(w_o + s)} \right] \qquad (4.29)$$

$$T(s) = \frac{(R_1 \tau_2 \tau_3 + R_2 \tau_1 \tau_3 + R_3 \tau_1 \tau_2)}{R_{in} \tau_1 \tau_2 \tau_3}$$

$$\left[\frac{\dfrac{R_1 + R_2 + R_3}{R_1 \tau_2 \tau_3 + R_2 \tau_1 \tau_3 + R_3 \tau_1 \tau_2} + s \left(\dfrac{R_1(\tau_2 + \tau_3) + R_2(\tau_1 + \tau_3) + R_3(\tau_1 + \tau_2)}{R_1 \tau_2 \tau_3 + R_2 \tau_1 \tau_3 + R_3 \tau_1 \tau_2} \right) + s^2}{\left(\dfrac{1}{\tau_1} + s \right) \left(\dfrac{1}{\tau_2} + s \right) \left(\dfrac{1}{\tau_3} + s \right)} \right]$$

$$(4.30)$$

$$w_{p1} = \frac{1}{\tau_1} = 2\pi f_{p1} = f_{p1} \Rightarrow 0.00206\,\text{Hz}$$

$$w_{p2} = \frac{1}{\tau_2} = 2\pi f_{p2} = f_{p2} \Rightarrow \quad 0.346\,\text{Hz} \qquad (4.31)$$

$$w_o = \frac{1}{\tau_3} = 2\pi f_o = f_{po} \Rightarrow \quad 57.87\,\text{Hz}$$

One can associate the values of the zeroes, which can be approached by expressions (4.32) and (4.33), where one can substitute the values of the circuit elements R and C from Table 4.2, and according to the approximations for the time constants τ, one gets (4.34).

$$w_{z1} \cdot w_{z2} = \frac{(R_1 + R_2 + R_3)}{(R_1\tau_2\tau_3 + R_2\tau_1\tau_3 + R_3\tau_1\tau_2)} \qquad (4.32)$$

$$w_{z1} + w_{z2} = \frac{R_1(\tau_2 + \tau_3) + R_2(\tau_1 + \tau_3) + R_3(\tau_1 + \tau_2)}{R_1\tau_2\tau_3 + R_2\tau_1\tau_3 + R_3\tau_1\tau_2} \qquad (4.33)$$

$$w_{z1} \cdot w_{z2} = 284.82 = A$$

$$w_{z1} + w_{z2} = 219.38 = B \qquad (4.34)$$

Performing algebraic operations one can solve for the angular frequencies by evaluating the quadratic equation: $w_{z2}^2 - Bw_{z2} + A = 0$, which is solved by applying the general quadratic formulae given in (4.35). In this manner, the frequencies of the zeroes are listed in (4.36). Also, one should evaluate the expression given in (4.37), which takes into consideration that $w_o = 1/\tau_3$, so that one obtains (4.38).

$$w_{z2} = \frac{-b \pm \sqrt{b^2 - 4ac}}{2a} = (0.3061, 218.075) \quad w_{z1} = A/w_{z2} = (218.25, 1.306)$$

$$(4.35)$$

$$w_{z1} = 2\pi f_{z1} = \quad 1.3061 = f_{z1} \Rightarrow 0.208\,\text{Hz}$$
$$w_{z2} = 2\pi f_{z2} = 218.075 = f_{z2} \Rightarrow 37.71\,\text{Hz} \qquad (4.36)$$

$$w_o G_{H1} G_{H2} G = \frac{R_1\tau_2\tau_3 + R_2\tau_1\tau_3 + R_3\tau_1\tau_2}{R_{in}\tau_1\tau_2\tau_3} \qquad (4.37)$$

$$G_{H1} G_{H2} G = \frac{R_1\tau_2\tau_3 + R_2\tau_1\tau_3 + R_3\tau_1\tau_2}{R_{in}\tau_1\tau_2} \Rightarrow 6.230 \qquad (4.38)$$

Table 4.6 shows the main parameters that are synthesized into the three CAMs in the FPAA, i.e., two pole/zero bilinear filters and one lowpass filter. These CAMs are connected in cascade as shown in Fig. 4.13, and Figs. 4.14, 4.15, and 4.16 show the configurations of the parameters that are used in the filter designs using the Anadigm Designer 2 EDA tool [19].

Table 4.6 Configurable Analog Modules in the FPAA for the implementation of the fractional-order integrator to approach $q = 0.9$ in the Lorenz system

CAM	Filter type	f_z	f_p	G_H
Bilinear 1	Pole zero filter	$f_{z1} = 0.208$ Hz	$f_{p1} = 0.00206$ Hz	$G_{H1} = 1.0$
Bilinear 2	Pole zero filter	$f_{z2} = 37.71$ Hz	$f_{p2} = 0.346$ Hz	$G_{H2} = 1.0$
Bilinear 3	Low pass filter	$f_o = 57.87$ Hz		$G_{H3} = 6.23$

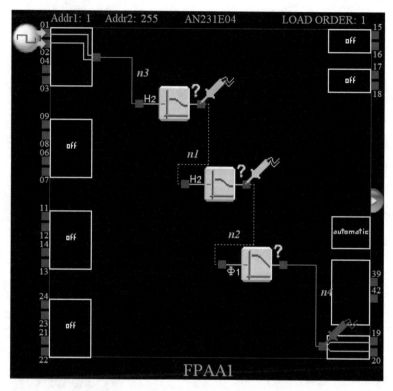

Fig. 4.13 Cascade connection of the bilinear pole/zero filters to approach the fractional order $q = 0.9$, using Anadigm Designer 2

There exists another topology for the implementation of the fractional-order integrator; for example, one can use a biquad filter block, which is directly available into the FPAA, plus a lowpass bilinear filter. However, the required resources in the FPAA will be the same as cascading three CAMs as described above. If one wants to use a biquadratic topology, it can be implemented using the transfer functions given in (4.21) and (4.23), which correspond to a biquad expression and a lowpass filter response. Afterwards, one should multiply the transfer functions as described in (4.39), which is the same symbolic transfer function as the one derived from Fig. 4.12, but by using a fractance with an RC-ladder topology.

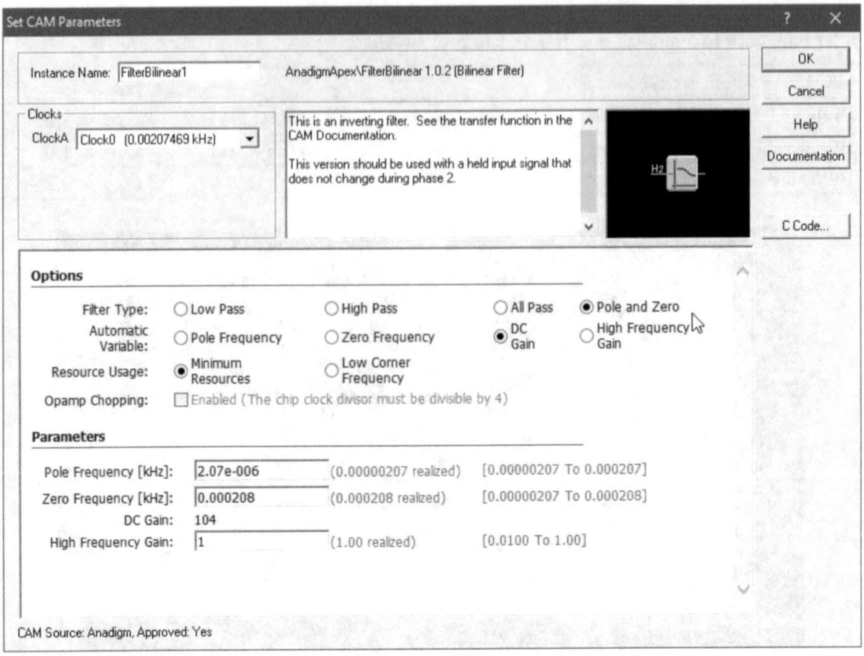

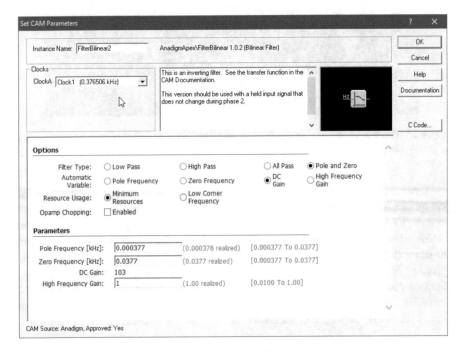

Fig. 4.14 CAM configuration of the first pole/zero bilinear filter

Fig. 4.15 CAM configuration of the second pole/zero bilinear filter

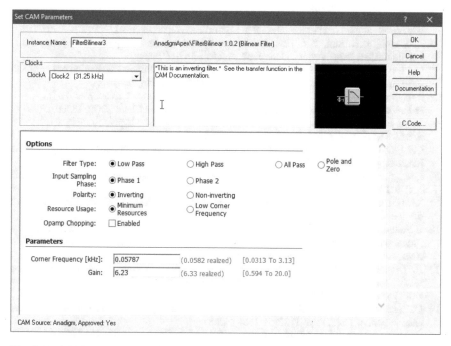

Fig. 4.16 CAM configuration of the third lowpass bilinear filter

$$T(s) = \frac{w_o G G_H \left(s^2 + \dfrac{w_z}{Q_z}s + w_z^2\right)}{s^3 + s^2\left(\dfrac{w_p}{q_p} + w_o\right) + s\left(w_p^2 + \dfrac{w_p w_o}{q_p}\right) + w_p^2 w_o} \tag{4.39}$$

The fractional-order integrator shown in Fig. 4.12 has three RC pairs in a ladder topology that perform the feedback in the amplifier. They give rise to three time constants RC having the values of 2.75 ms, 0.46 s, and 77.3 s, respectively. These values can be scaled in frequency to reach higher dynamics of the fractional-order chaotic oscillators, and to exploit the FPAA resources. The process of frequency scaling using integer-order integrators can be found in active filter design books.

References

1. C. Li, J. Yan, The synchronization of three fractional differential systems. Chaos Solitons Fractals **32**(2), 751–757 (2007)
2. H.Y. Jia, Q. Tao, Z.Q. Chen, Analysis and circuit design of a fractional-order Lorenz system with different fractional orders. Syst. Sci. Control Eng. Open Access J. **2**(1), 745–750 (2014)
3. C. Li, J. Yan, The synchronization of three fractional differential systems. Chaos Solitons Fractals **32**(2), 751–757 (2007)

4. H.Y. Jia, Q. Tao, Z.Q. Chen, Analysis and circuit design of a fractional-order Lorenz system with different fractional orders. Syst. Sci. Control Eng. Open Access J. **2**(1), 745–750 (2014)
5. A. Charef, H.H. Sun, Y.Y. Tsao, B. Onaral, Fractal system as represented by singularity function. IEEE Trans. Autom. Control **37**(9), 1465–1470 (1992)
6. L. Chen, W. Pan, R. Wu, K. Wang, Y. He, Generation and circuit implementation of fractional-order multi-scroll attractors. Chaos Solitons Fractals **85**, 22–31 (2016)
7. S.-P. Wang, S.-K. Lao, H.-K. Chen, J.-H. Chen, S.-Y. Chen, Implementation of the fractional-order Chen–Lee system by electronic circuit. Int. J. Bifurcation Chaos **23**(2), 1350030 (2013)
8. Y. Yu, H.-X. Li, The synchronization of fractional-order Rössler hyperchaotic systems. Phys. A Stat. Mech. Appl. **387**(5–6), 1393–1403 (2008)
9. C. Li, G. Chen, Chaos and hyperchaos in the fractional-order rössler equations. Phys. A Stat. Mech. Appl. **341**, 55–61 (2004)
10. I. Grigorenko, E. Grigorenko, Chaotic dynamics of the fractional Lorenz system. Phys. Rev. Lett. **91**(3), 034101 (2003)
11. W.M. Ahmad, J.C. Sprott, Chaos in fractional-order autonomous nonlinear systems. Chaos Solitons Fractals **16**(2), 339–351 (2003)
12. I. Podlubny, *Fractional Differential Equations: An Introduction to Fractional Derivatives, Fractional Differential Equations, to Methods of Their Solution and Some of Their Applications*, vol. 198 (Elsevier, Amsterdam, 1998)
13. A. Djouambi, A. Charef, A. BesançOn, Optimal approximation, simulation and analog realization of the fundamental fractional order transfer function. Int. J. Appl. Math. Comput. Sci. **17**(4), 455–462 (2007)
14. C. Xiang-Rong, L. Chong-Xin, W. Fa-Qiang, Circuit realization of the fractional-order unified chaotic system. Chin. Phys. B **17**(5), 1664 (2008)
15. C. Muñiz-Montero, L.V. García-Jiménez, L.A. Sánchez-Gaspariano, C. Sánchez-López, V.R. González-Díaz, E. Tlelo-Cuautle, New alternatives for analog implementation of fractional-order integrators, differentiators and PID controllers based on integer-order integrators. Nonlinear Dyn. **90**(1), 241–256 (2017)
16. C. Muñiz-Montero, L.A. Sánchez-Gaspariano, C. Sánchez-López, V.R. González-Díaz, E. Tlelo-Cuautle, On the electronic realizations of fractional-order phase-lead-lag compensators with OpAmps and FPAAs, in *Fractional Order Control and Synchronization of Chaotic Systems* (Springer, Berlin, 2017), pp. 131–164
17. J. Lü, G. Chen, Generating multiscroll chaotic attractors: theories, methods and applications. Int. J. Bifurcation Chaos **16**(4), 775–858 (2006)
18. T.J. Freeborn, B. Maundy, A.S. Elwakil, Field programmable analogue array implementation of fractional step filters. IET Circuits, Devices Syst. **4**(6), 514–524 (2010)
19. Anadigm, Dynamically Reconfigurable dpASP, 3rd Generation, AN231E04 Datasheet Rev 1.2 (2014). www.anadigm.com

Chapter 5
FPGA-Based Implementations of Fractional-Order Chaotic Systems

5.1 On Grünwald-Letnikov Method

Nowadays, the number of applications of fractional calculus rapidly grows. This mathematical phenomenon allows us to describe and model a real object more accurately than it has been done applying classical integer methods. In the literature one can find many numerical methods for the approximation of the fractional-order derivative and integral, advising that fractional calculus can easily be used in very wide areas of applications. In fact, fractional calculus is a generalization of integration and differentiation operations, and it proposes the fundamental operator $_aD_t^\alpha$, where α and t are the bounds of the operation and $\alpha \in R$. The continuous integro-differential operator is defined as:

$$a = \begin{cases} \frac{d^\alpha}{dt^\alpha}, & \alpha > 0 \\ 1, & \alpha = 0 \\ \int_\alpha^t (d\tau)^\alpha, & \alpha < 0 \end{cases} \qquad (5.1)$$

For performing the numerical simulation of fractional-order chaotic oscillators, one can find various numerical methods; however, this chapter focuses on Grünwald-Letnikov's definition and Adams-Bashforth-Moulton predictor-corrector scheme [1].

As described in Chap. 1, considering the continuous function $f(t)$, its first, second, and third derivatives can be expressed as [2]:

$$\frac{d}{dt} = f'(t) = \lim_{h \to 0} \frac{f(t) - f(t-h)}{h} \qquad (5.2)$$

$$\frac{d^2}{dt^2} = f''(t) = \lim_{h \to 0} \frac{f(t) - 2f(t-h) + f(t-2h)}{h^2} \qquad (5.3)$$

© Springer Nature Switzerland AG 2020
E. Tlelo-Cuautle et al., *Analog/Digital Implementation of Fractional Order Chaotic Circuits and Applications*, https://doi.org/10.1007/978-3-030-31250-3_5

$$\frac{d^3}{dt^3} = f'''(t) = \lim_{h \to 0} \frac{f(t) - 3f(t - h) + 3f(t - 2h) - f(t - 3h)}{h^3} \tag{5.4}$$

One can observe a pattern, so that the n-derivative of $f(t)$ for $n \in N$, $j > n$ can be described by (5.5). Henceforth, according to (5.2)–(5.5) one can write the fractional-order derivative definition of order α ($\alpha \in R$), as given in (5.6).

$$\frac{d^n}{dt^n} f(t) = \lim_{h \to 0} \frac{1}{h^n} \sum_{j=0}^{n} (-1)^j j \binom{n}{j} f(t - jh) \tag{5.5}$$

$$_a D_t^\alpha f(t) = \lim_{h \to 0} \frac{_a \Delta_h^\alpha f(t)}{h^\alpha}, \quad _a \Delta_h^\alpha f(t) = \sum_{j=0}^{\frac{t-a}{h}} (-1)^j \binom{\alpha}{j} f(t - jh) \tag{5.6}$$

In this manner, the general form of a fractional-order chaotic oscillator having three equations and applying the Grünwald-Letnikov definition given in (5.6) leads us to the system of equations given in (5.7), where α is any positive real number, but for fractional-order chaotic oscillators it is between $0 < \alpha < 1$, h is the step-size and $c_j^{(\alpha)}$ denotes the binomial coefficients.

$$x(t_k) = f(x(t_k), t_k) h^{\alpha_1} - \sum_{j=v}^{k} c_j^{(\alpha_1)} x(t_{k-j})$$

$$y(t_k) = f(y(t_k), t_k) h^{\alpha_2} - \sum_{j=v}^{k} c_j^{(\alpha_2)} y(t_{k-j}) \tag{5.7}$$

$$z(t_k) = f(z(t_k), t_k) h^{\alpha_3} - \sum_{j=v}^{k} c_j^{(\alpha_3)} z(t_{k-j})$$

That way, the implementation of the fractional difference method to compute fractional derivatives requires the evaluation of the binomial coefficients, which are computed using (5.8), from which one can obtain the recurrent relationships given in (5.9).

$$c_j^\alpha = (-1)^\alpha \binom{\alpha}{k}, \quad j = 0, 1, 2, \ldots \tag{5.8}$$

$$c_0^{(\alpha)} = 1, \quad c_j^{(\alpha)} = \left(1 - \frac{1 + \alpha}{j}\right) c_{j-1}^{(\alpha)} \tag{5.9}$$

Using the equations given above is suitable for problems setting a fixed value of α. It allows the creation of an array of coefficients which can be used for fractional differentiation of various functions. For example, Table 5.1 and Fig. 5.14 show the relationship between the values of c_j^α and the index j for different fractional-orders α, when $0 < \alpha < 1$. As one can infer, in this case it can be concluded that as index j increases, the magnitude of the binomial coefficients tends to zero. Henceforth, the dependence of the values of the function decreases in the same rate as α approaches to an integer value [3].

For $t \gg a$ the number of addends in the fractional-derivative approximation (5.6) becomes enormously large, then one needs more and more terms to add for computing the final solution. In other words, the final solution will need unlimited memory. Therefore, it is necessary to include a formulation related to reduce the memory requirements to store the binomial coefficients, and this is well known as "short memory" principle, which means taking into account the behavior of $f(t)$ only in the few recent past values. The term memory expressed by the sum in (5.7) is the "short memory" and it allows to retain a limited amount of information for a relatively short period of time. Then the lower index of the sum in (5.7) will be defined by (5.10), where $k = 1, 2, 3, 4, \ldots$ and L is the memory length. Therefore, as shown in Fig. 5.1, the number of binomial coefficients is equal to m, and as k increases the memory locations being filling; once it is full, that is $k > m$, only the previous m-values of $x_k, x_{k-1}, \ldots, x_{k-m+1}$ are used by the summation.

$$v = \begin{cases} 1 & \text{when } k \leq m \\ k - m & \text{when } k > m \end{cases}, \qquad m = \frac{L}{h} \qquad (5.10)$$

Due to this approximation, the number of addends in (5.7) is always no greater than m. Of course, for this simplification one must pay a penalty in the form of some inaccuracy. For instance, Fig. 5.2 shows a graphical representation of the behavior of the binomial coefficients varying α. In this book, the short-memory principle is applied to implement several fractional-order chaotic oscillators.

Table 5.2 shows the equations, design parameters, initial conditions, fractional-order of the derivatives, and step size of five fractional-order chaotic oscillators. Those chaotic oscillators are called: Lorenz, Rössler, Chen, Lü, and V-Shape, which are simulated herein applying the definition of Grünwald-Letnikov (5.7). The numerical simulations to observe the fractional-order chaotic attractors are shown in Fig. 5.3 for Lorenz, Fig. 5.4 for Rössler, Fig. 5.5 for Chen, Fig. 5.6 for Lü, and Fig. 5.7 for V-Shape.

In the following sections the effect of the short memory principle and the implementation of these fractional-order chaotic oscillators on FPGAs will be shown. It will then be shown that the experimental fractional-order chaotic attractors, observed in an oscilloscope, have the same characteristics as those simulated herein.

Table 5.1 Values of c_j^α for $\alpha = 0.92$, $\alpha = 0.93$, and $\alpha = 0.99$

j	1	2	3	4	5	6	7	8	9
$\alpha = 0.92$	−0.92	−0.0368	−0.0132	−0.0069	−0.0042	−0.0029	−0.0021	−0.0016	−0.0013
$\alpha = 0.93$	−0.93	−0.0325	−0.0116	−0.0060	−0.0037	−0.0025	−0.0018	−0.0014	−0.0011
$\alpha = 0.99$	−0.99	−0.0050 −0.0017	−8.374e−4	−5.041e−4	−3.369e−4	−2.411e−4	−1.811e−4	−1.411e−4	

$$
\left.
\begin{aligned}
&y_0 c_0 \\
&y_1 c_0 + y_0 c_1 \\
&y_2 c_0 + y_1 c_1 + y_0 c_2 \\
&y_3 c_0 + y_2 c_1 + y_1 c_2 + y_0 c_3 \\
&y_4 c_0 + y_3 c_1 + y_2 c_2 + y_1 c_3 + y_0 c_4 \\
&\qquad\qquad\qquad \vdots \\
&y_k c_0 + y_{k-1} c_1 + y_{k-2} c_2 + y_{k-3} c_3 + \dots + y_2 c_{m-2} + y_1 c_{m-1} + y_0 c_m
\end{aligned}
\right\} k \le m
$$

$$
\left.
\begin{aligned}
&y_{k+1} c_0 + y_k c_1 + y_{k-1} c_2 + y_{k-2} c_3 + \dots + y_3 c_{m-2} + y_2 c_{m-1} + y_1 c_m \\
&y_{k+2} c_0 + y_{k+1} c_1 + y_k c_2 + y_{k-1} c_3 + \dots + y_4 c_{m-2} + y_3 c_{m-1} + y_2 c_m \\
&\qquad\qquad\qquad \vdots
\end{aligned}
\right\} k > m
$$

Fig. 5.1 Short memory principle expressed by (5.7)

5.2 Effects of the Short Memory Principle

As described above, the simulation of fractional-order chaotic oscillators requires a huge number of memory locations to accumulate the coefficients and then guarantee minimum error with respect to the best approximation. In the ideal case, the truth solution will need an infinity number of memory locations. Is for this reason that one should approximate the solution with a numerical method that can reduce the memory requirements. In the majority of cases, the simulation can be performed applying the short-memory principle. This fact allows us to approximate the numerical solution by using the information of the "recent past," or in other words, the data saved in the interval $[t - L, t]$, with L being the length of memory. This is associated with a moving of a low limit to compute the derivatives [4], which are described by (5.11), in which the number of terms to add is limited by the value of L/h. The error of the approximation when $|f(t)| \ll M, (0 < t \ll t_1)$ is then bounded by (5.12), which can be used to determine the necessary memory length to obtain a certain error bound.

$$
D^\alpha f(t) \approx_{t-L} D^\alpha f(t), \quad t > L \tag{5.11}
$$

$$
\epsilon(t) = |D^\alpha f(t) -_{t-L} D^\alpha f(t)| \ll \left(\frac{M L^{-\alpha}}{|\Gamma(1-\alpha)|} \right)^{1/\alpha} \tag{5.12}
$$

The selection of the step-size h and the length of the memory L affect the numerical solution of the fractional-order chaotic oscillators. For example, Table 5.3 shows different solutions to observe the fractional-order chaotic attractor of Rössler oscillator when using different memory length (L) and step-sizes (h). As one can observe, the dynamics of the chaotic attractor decrease as h decreases under fixed L. In addition, when L is small enough, as h decreases the chaotic behavior tends to

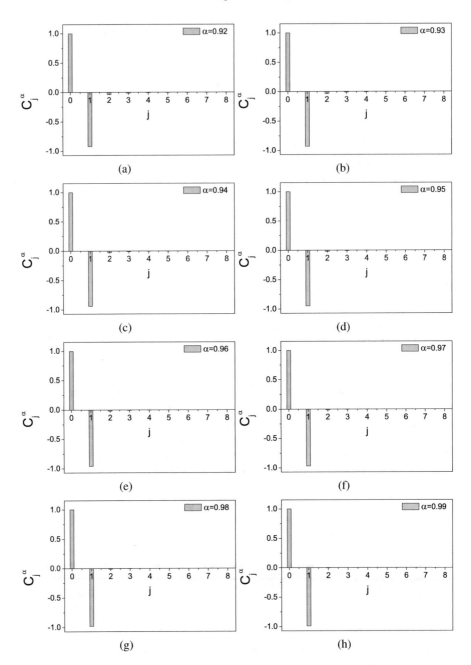

Fig. 5.2 Graphical representation of the behavior of the binomial coefficients c_j^{α} for (**a**) $\alpha = 0.92$, (**b**) $\alpha = 0.93$, (**c**) $\alpha = 0.94$, (**d**) $\alpha = 0.95$, (**e**) $\alpha = 0.96$, (**f**) $\alpha = 0.97$, (**g**) $\alpha = 0.98$, and (**h**) $\alpha = 0.99$

Table 5.2 Five fractional-order chaotic oscillators and their simulation conditions

System	Equations	Parameters	Initial conditions (x_0, y_0, z_0)	Fractional order $(\alpha_1, \alpha_2, \alpha_3)$	Step size (h)
Lorenz	$D^{\alpha_1} x = \sigma(y - x)$ $D^{\alpha_2} y = x(\rho - z) - y$ $D^{\alpha_3} z = xy - \beta z$	$\sigma = 10, \rho = 28,$ $\beta = 8/3$	$(0.1, 0.1, 0.1)$	$(0.994, 0.994, 0.994)$	0.005
Rössler	$D^{\alpha_1} x = -y - z$ $D^{\alpha_2} y = x + ay$ $D^{\alpha_3} z = b + z(x - c)$	$a = 0.5, b =$ $0.2, c = 10$	$(0.5, 0.5, 0.5)$	$(0.9, 0.9, 0.9)$	0.005
Chen	$D^{\alpha_1} x = a(y - x)$ $D^{\alpha_2} y = (c - a)x$ $\quad\quad -xz + cy$ $D^{\alpha_3} z = xy - bz$	$a = 35, b =$ $3, c = 28$	$(-9, -5, 14)$	$(0.9, 0.9, 0.9)$	0.005
Lü	$D^{\alpha_1} x = a(y - x)$ $D^{\alpha_2} y = -xz + cy$ $D^{\alpha_3} z = xy - bz$	$a = 36, b =$ $3, c = 20$	$(0.2, 0.5, 0.3)$	$(0.95, 0.95, 0.95)$	0.005
H-Shape	$D^{\alpha_1} x = y - x$ $D^{\alpha_2} y = sign(x)$ $\quad\quad [1 - mz + (Gz)]$ $D^{\alpha_3} z = |x| - rz$	$m = 2, r = 0.5$	$(0.5, 1, 0.5)$	$(0.97, 0.97, 0.97)$	0.05

disappear. On the other hand, for the simulations with fixed step-size h, the chaotic behavior increases as L increases, which sounds logic due to the span of more signal history of the fractional-order operator [5]. Table 5.4 shows different numerical solutions to observe the fractional-order chaotic attractor of V-Shape oscillator, also when using different L and h. As one can observe, the h is quite different than that for Rössler and also for the other fractional-order chaotic oscillators listed in Table 5.2. Different to Table 5.3, the dynamics of V-Shape chaotic attractor require more memory length to be observed. Besides also the dynamical behavior decreases as h decreases under fixed L, and if L is small, as h decreases the chaotic behavior tends to disappear. Finally, for a fixed step size h, the chaotic behavior increases as L increases.

The Grünwald-Letnikov method can be implemented on embedded systems like FPGAs, which have experienced a rapid growth in industry and have been a viable alternative infrastructure for Application-Specific Integrated Circuits (ASICs). Nowadays, FPGAs have become an indispensable part of digital systems and have widespread usage in a wide domain of digital systems [6].

Xilinx and Altera are the main leading FPGA manufacturers. Besides, each FPGA vendor designs an own reconfigurable architecture. But, in most cases, FPGAs consist of three basic blocks that are configurable logic blocks (CLBs), in-out blocks (I/O ports), and connection blocks. Figure 5.8 shows a general structure of a FPGA where one can see the basic blocks. New FPGAs been equipped with additional hardware components such as memory blocks (single/dual

Fig. 5.3 Phase-space portraits of Lorenz fractional-order chaotic oscillator applying Grünwald-Letnikov method, and plotting two state variables to observe: (**a**) X–Y view, (**b**) X–Z view, and (**c**) Y–Z view

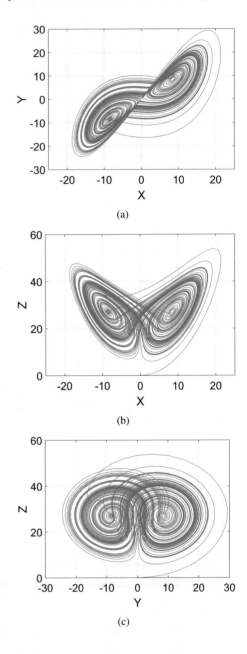

(a)

(b)

(c)

port RAMs), embedded multipliers, and Digital Signal Processors (DSPs), and intellectual property (IP) cores. These are optimally designed and embedded in the devices to facilitate the implementation of specific functions.

The design of a digital circuit on an FPGA can be performed from the description of the hardware using a hardware description language (HDL). In this case, there are

Fig. 5.4 Phase-space portraits of Rössler fractional-order chaotic oscillator applying Grünwald-Letnikov method, and plotting two state variables to observe: (**a**) X–Y view, (**b**) X–Z view, and (**c**) Y–Z view

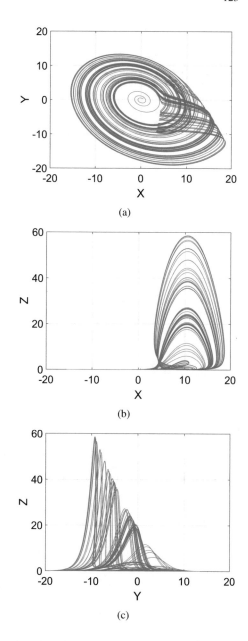

(a)

(b)

(c)

two IEEE (Institute of Electrical and Electronics Engineers) standards that are in use in industry and academia, namely: Verilog and VHDL. Both languages are suitable for the majority of FPGAs, and this chapter is devoted to apply VHDL to describe architectures as shown in Chap. 2.

Fig. 5.5 Phase-space
portraits of Chen
fractional-order chaotic
oscillator applying
Grünwald-Letnikov method,
and plotting two state
variables to observe: (**a**) X–Y
view, (**b**) X–Z view, and (**c**)
Y–Z view

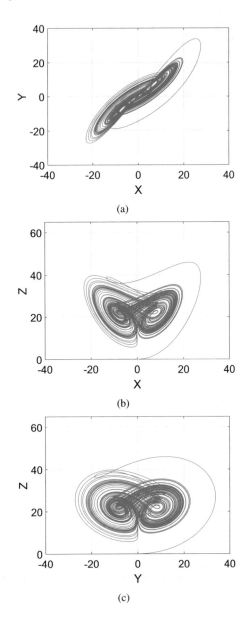

(a)

(b)

(c)

5.3 Hardware Descriptions in VHDL

The general structure of a hardware description under VHDL consists of a set of
units that define, describe, structure, and analyze a digital system. There are three
fundamental units in VHDL: Library and package, Entity, and Architecture.

Fig. 5.6 Phase-space portraits of Lü fractional-order chaotic oscillator applying Grünwald-Letnikov method, and plotting two state variables to observe: (**a**) X–Y view, (**b**) X–Z view, and (**c**) Y–Z view

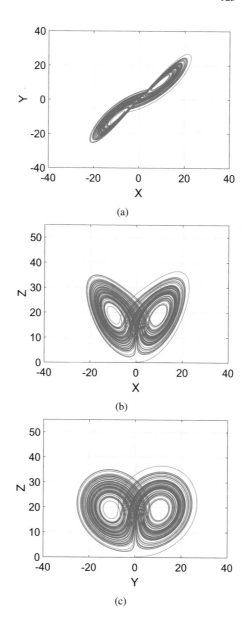

(a)

(b)

(c)

5.3.1   *Library Package*

The VHDL language is standardized by IEEE, who created the IEEE VHDL library and std_logic type in standard 1164. In this manner, all the descriptions under VHDL will begin with the header of libraries. One example is given in Listing 5.1, which allows the use of all the digital components included in library ieee

Fig. 5.7 Phase-space
portraits of V-Shape
fractional-order chaotic
oscillator applying
Grünwald-Letnikov method,
and plotting two state
variables to observe: (**a**) X–Y
view, (**b**) X–Z view, and (**c**)
Y–Z view

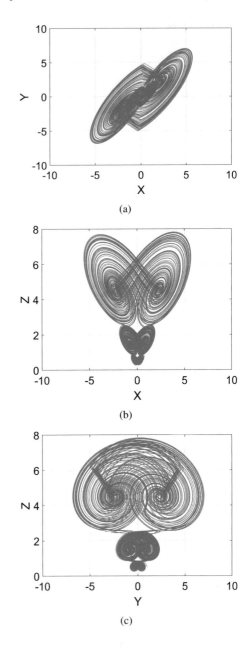

(a)

(b)

(c)

and the use of the package `std_logic_1164` that includes the type of data
(information being transmitted by a port) usually required in VHDL descriptions
(see Table 5.5).

Table 5.3 Simulation of Rössler fractional-order oscillator applying Grünwald-Letnikov with different memory length L and step-size h

	$h = 0.005$	$h = 0.01$	$h = 0.05$
$L = 16$			
$L = 32$			
$L = 64$			
$L = 128$			
$L = 256$			
$L = 512$			

Table 5.4 Simulation of V-Shape fractional-order oscillator applying Grünwald-Letnikov with different memory length L and step-size h

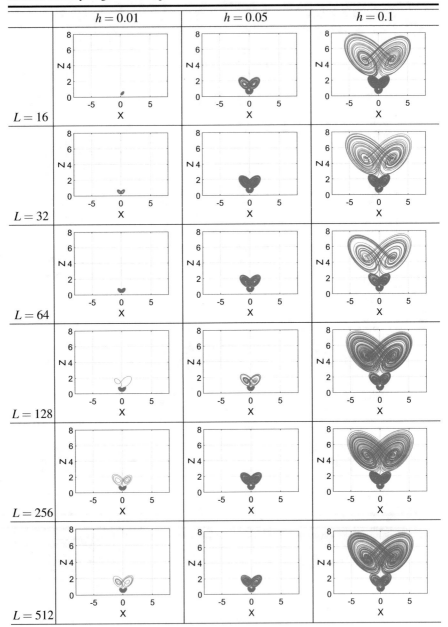

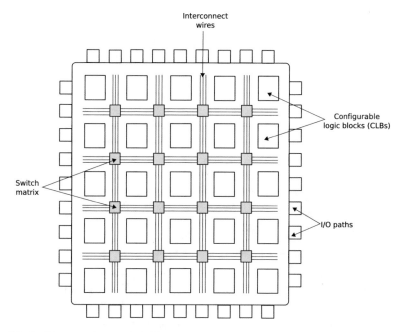

Fig. 5.8 Architecture of a generic FPGA

Table 5.5 Types of data used in VHDL

Data type	Description
std_logic	Has 9 possible values (U, X, 0,1,Z, W, L, H,–)
std_logic_vector	Vector of bits of type std_logic
Integer	Represents an integer number
Bit	Takes the value of 0 or 1
bit_vector	Vector of bits

```
1   library ieee;
2   use ieee.std_logic_1164.all;
```

Listing 5.1 Library package

In case of performing mathematical operations within an FPGA, it will be necessary to add a special package to develop this task when describing under VHDL, it is the library ieee.numeric_std, which allows performing arithmetic functions for vectors. This package defines two numeric types:

Table 5.6 Arithmetic
operators commonly used in
VHDL

Operation	Operator	Example
Addition	+	C<=A+B;
Subtraction	−	C<=A−B;
Multiplication	*	C<=A*B
Exponentiating	**	C<=2**2;
Absolute value	abs	C<= abs(A);
Remainder	rem	C<= a rem b;
Modulus	mod	C<= A mod B;

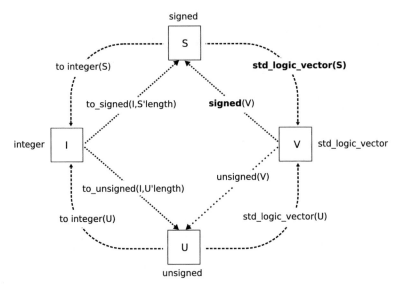

Fig. 5.9 Useful data type conversion functions

- Unsigned. Represents an unsigned number in vector form
- Signed. Represents a signed number in vector form and in two's complement.

In addition, the package `ieee.numeric_std` contains the arithmetic operators listed in Table 5.6, which includes the signed and unsigned types, and also it
contains useful type conversion functions like the ones shown in Fig. 5.9.

5.3.2   Entity Declaration

The declaration of an entity is generally visualized as a black box, where is
performed the description of the inputs and outputs of the digital block at any level
of abstraction. In this case those terminals are associated with a direction of type

Table 5.7 Direction of the information of the input and output ports

Mode	Description
in	Input signals of an entity (unidirectional)
out	Output signals of an entity (unidirectional)
inout	Allows feedback of the signals in or out of the entity (bidirectional)
buffer	Similar to mode inout, but behaves as an output port or terminal

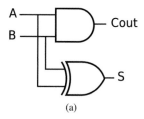

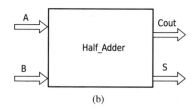

(a)   (b)

Fig. 5.10 Half adder: (**a**) internal structure and (**b**) entity as a black box description

input (in) and output (out). All ports declared in the entity must have a name, mode, and type of data from Table 5.5. The types of mode used in VHDL are given in Table 5.7.

Listing 5.2 shows the entity description of a half adder, which architecture is shown in Fig. 5.10a. As a black box the abstraction is shown in Fig. 5.10b, where it can be appreciated that the ports A and B are of type `in` and the ports Cout and S are of type `out`. It is worth mentioning that `entity` is a reserved word in VHDL. The entity is declared and followed by its name, in this case: `Half_Adder`, and afterwards one should place the reserved word `is`. The input and output ports are declared within the instruction `port`, in this case, and as shown in Fig. 5.10, A and B have mode `in`, while the output ports `Cout` and `S` have mode `out`. The types of data being used are in the library `std_logic`. The entity declaration ends with the reserved word `end`, followed by the name of the entity `Half_Adder`.

```
1   entity Half_Adder is
2       port(
3           A, B : in std_logic;
4           Cout, S : out std_logic
5           );
6   end Half_Adder;
```

Listing 5.2 Entity description of a half adder

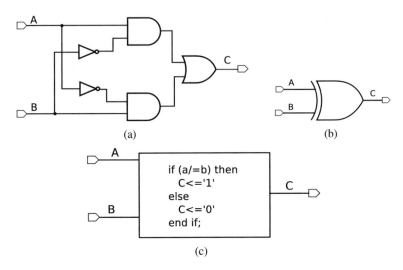

(a)

(b)

(c)

Fig. 5.11 Programming styles for the design of an architecture: (**a**) Data flow, (**b**) Structural, and (**c**) Behavioral

5.3.3 Architecture

The entity is used to declare the black box of a digital block, and the architecture is used to describe the behavior of the entity to accomplish the desired function. The styles of programming to describe the architecture of an entity can be classified into three types:

- Data flow: This style shows in detail the transfer of information from the input to the output.
- Structural: This style describes the black box using logic models already established, e.g., using adders, subtractors, multipliers, logic gates, etc. This description style can help to organize a projects in a hierarchical fashion.
- Behavioral. This style requires knowing the behavior of the entity. That way, it is only required to know the behavior of the outputs with respect to the inputs.

The three programming styles are shown in Fig. 5.11, all the cases describe the behavior of an XOR gate, which truth table basically indicates that the output C is equal to a logic 1 when the inputs A and B are equal, otherwise, the output C equals to a logic 0. Listing 5.3 shows the entity description of the XOR gate shown in Fig. 5.11a, which is implemented with NOT, AND, and OR gates. In this case, one can note that a VHDL description based on the data flow will require that the instructions be executed in a concurrent way. This type of description allows defining a flow of information from one gate to another.

```
1   library ieee;
2   use ieee.std_logic_1164.all;
3
4   entity XOR_dataflow is
5     port(
6     a, b: in std_logic;
7     c: out std_logic);
8   end XOR_dataflow;
9
10  Architecture behavioral of XOR_dataflow is
11  begin
12        C<=(A and(not B)) or (B and (not A));
13  end behavioral;
```

Listing 5.3 Description of an architecture based on a data flow style

Listing 5.4 shows the entity description of the XOR gate shown in Fig. 5.11b, and using a structural description, which is based on component instantiation statements.

```
1   library ieee;
2   use ieee.std_logic_1164.all;
3
4   entity XOR_structural is
5     port(
6     a, b: in std_logic;
7     c: out std_logic);
8   end XOR_structural;
9
10  Architecture behavioral of XOR_structural is
11  component Gate_xor
12      port(
13        a, b: in std_logic;
14        c: out std_logic
15      );
16  end component;
17  begin
18      BQ1: Gate_xor port map(a,b,c);
19  end behavioral;
```

Listing 5.4 Description of an architecture in an structural style

Listing 5.5 shows the entity description of the XOR gate shown in Fig. 5.11c, in which the behavioral description is appreciated, where the architecture uses only process statements. Recall that VHDL is a concurrent language, so that it does not follow an order in which the instruction are written. That way, if two instructions exist, both can be executed at the same time. However, in this case should be necessary the use of a sequential declaration, where the instructions must be executed in the order that they appear in the written code. In this case one can use the instruction process, which will have assigned a list of sensitivity of the variables of interest.

```
1   library ieee;
2   use ieee.std_logic_1164.all;
3
4   entity XOR_behavioral is
5     port(
6     a, b: in std_logic;
7     c: out std_logic);
8   end XOR_behavioral;
9
10  Architecture behavioral of XOR_behavioral is
11  begin
```

```
12   process (a,b)
13   begin
14      if (a/=b) then
15         C<='1';
16      else
17         C<='0';
18      end if;
19   end process;
20   end behavioral;
```

Listing 5.5 Description of an architecture based on a behavioral style

Going back to example of the half adder shown in Fig. 5.10, Listing 5.6 shows its architecture description. As one can observe in the examples described above, the architecture is programmed using the reserved word `architecture` followed by an arbitrary name to identify it, and in this case is `behavioral`, and also one must include the entity that is associated with `Half_Adder`. In the description it can be noted the reserved word `begin`, which is used to indicate the beginning to declare the processes that are part of the behavior of the whole entity.

```
1   -- Architecture declaration of a half adder
2   architecture behavioral of Half_Adder is
3   begin
4      S    <= A xor B;
5      Cout <= A and B;
6   end behavioral;
```

Listing 5.6 Entity description of a half adder

5.4 Computer Arithmetic

All kind of digital hardware must process data that is represented in a specific way. In the case of an FPGA, it allows a wide variety of computer arithmetic implementations for the desired digital signal processing algorithms, because of the physical bit-level programming architecture [7]. The data representation can be represented in two groups: fixed-point and floating point notation. In fixed-point a given number has a specific number of bits, which are reserved for the sign, the integer part, and the fractional part. In the case of floating point numbers, they are similar to the scientific notation, including a sign, mantissa, and an exponent.

When operating on data with large dynamic range it is a good option using the floating point representation. However, implementing floating-point arithmetic operations on FPGAs is very expensive in terms of the required number of logic elements. Also, the arithmetic operations such as multiplication, and even more the division, can be costly if used too freely. For these reasons, a careful hardware design is needed to minimize the number of these operations [8].

The fixed-point representation of real numbers has higher speed and lower cost of hardware implementation than using floating point notation. For these reasons this chapter uses fixed-point representation for the digital synthesis of fractional-order chaotic oscillators. The signed fixed-point numbers are also represented as two's complement ones. In this case, from a fixed-point representation, the two's complement numbers are formed by inverting the bits of the absolute value and adding a logic 1 to the least significant bit.

5.4.1 Example: Fixed-Point Representation of −2.75

This part details the steps to be followed to implement the binary representation of the real number −2.75, using the fixed-point format and two's complement. In this case, one bit will be used to represent the sign, four bits to represent the integer part, and nine bits for the fractional part.

- Step 1: Represent | −2.75 | in binary format using one bit for the sign, four bits for the integer part, and nine bits for the fractional part, as shown in Fig. 5.12. A binary point must be added between the bits representing the integer and fractional parts, and this is analogous to the decimal point. Besides, this point does not exist in practice and the designer is the unique person who knows its location.
- Step 2: Invert the bits from step 1 and add a logic 1 to the least significant bit. The result of the sum is the real number −2.75 expressed in two's complement representation, as shown in Fig. 5.13.

In this chapter, the majority of fixed-point representations are done using 32 bit in total, where the number of bits in the integer and fractional parts will depend on

Sign		Integer part			Fractional part								
		2^3 2^2 2^1 2^0			2^{-1}	2^{-2}	2^{-3}	2^{-4}	2^{-5}	2^{-6}	2^{-7}	2^{-8}	2^{-9}
0		0 0 1 0	.		1	1	0	0	0	0	0	0	0

Fig. 5.12 Binary representation of the real number | −2.75 | using one bit for the sign, four bits for the integer part, and nine bits for the fractional part

Sign		Integer part			Fractional part								
		2^3 2^2 2^1 2^0			2^{-1}	2^{-2}	2^{-3}	2^{-4}	2^{-5}	2^{-6}	2^{-7}	2^{-8}	2^{-9}
1		1 1 0 1	.		0	0	1	1	1	1	1	1	1
0		0 0 0 0	.		0	0	0	0	0	0	0	0	1
1		1 1 0 1	.		0	1	0	0	0	0	0	0	0

Fig. 5.13 Representing −2.75 in fixed-point and expressed in two's complement. It is obtained after inverting the bits of the absolute value from Fig. 5.12, and adding a logic 1 to the least significant bit

the ranges of the amplitudes of the state variables of the fractional-order chaotic oscillator being implemented on FPGA.

5.5   VHDL Description of Some Most Used Digital Blocks

In Chap. 2 the main digital blocks known as adder, subtractor, multiplier, and counter were described under VHDL. In this chapter other blocks and some of the previous ones are described with the aim to be used within the description of a whole fractional-order chaotic oscillator.

5.5.1   Adder and Subtractor

In the FPGA-based implementation of fractional-order chaotic oscillators, the basic digital block is the adder, from which one can describe the subtractor, and both blocks can work with two numbers represented by N number of bits. One can describe asynchronous and synchronous blocks, but the former ones may lead to problem in synchronizing the operations during the iterations for solving equations that have different number of operations and then require different number of clock cycles or they need to be reseted. In this manner, the arithmetic blocks that are used herein are of synchronous type, so that they include one pin for controlling the clock CLK and another pin for the reset RST operation. These pins are connected to registers, usually to flip-flops that store the data as a result of the sum or subtraction operation. That way, the synchronous blocks are controlled to store the right data at each iteration because wrong data might be written into the memory elements [9].

In the case of the subtraction operation to evaluate $S = A - B$, this operation can be performed through creating the two's complement representation of B to obtain $-B$, and afterwards one can program the addition operation between $A + (-B)$. Figure 5.14a shows the entity of the block subtractor, and in Fig. 5.14b its internal architecture is shown. In both cases one can see that the length of the data buses have 32 bits, while it is required one bit for the clock CLK and one bit for the reset RST.

Figure 5.15a shows the entity of the adder block, which can sum either positive or negative numbers when using the two's complement representation, i.e., $S = A + B$. The architecture or synthesis is shown in Fig. 5.15b.

Listing 5.7 shows the VHDL description of a 32 bit adder with fixed-point format. The addition operation is performed as follows: after doing a reset, the 32 bit addends are loaded and the sum register stores the result that is transferred to the output register. This VHDL code converts data types using the instruction std_logic_vector to get a signed data or vice versa, so that one uses the conversion shown in Fig. 5.9. The VHDL code for a subtractor is quite similar, but in this case one must update the name of the entity and replace the symbol $+$ by $-$.

Fig. 5.14 Subtractor: (**a**) Black box description or entity, and (**b**) digital hardware description or synthesis

Fig. 5.15 Adder: (**a**) Black box description or entity, and (**b**) digital hardware description or synthesis

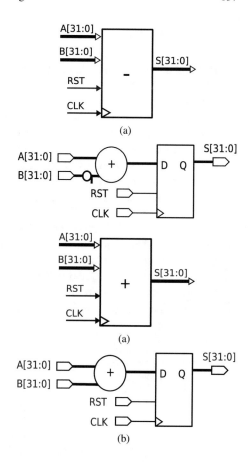

```
1   library ieee;
2   use ieee.std_logic_1164.all;
3   use ieee.numeric_std.all;
4
5   entity Adder is
6       port(
7           CLK, RST: in std_logic;
8           A, B    : in std_logic_vector(31 downto 0);
9           Output  : out std_logic_vector(31 downto 0) :=(others => '0'));
10  end Adder;
11
12  architecture behavioral of Adder is
13      begin
14          process (CLK, RST) begin
15              if RST = '0' then
16                  Output <= (others => '0');
17              elsif rising_edge(CLK)  then
18          Output <= std_logic_vector(signed(A) + signed(B));
19              end if;
20          end process;
21  end behavioral;
```

Listing 5.7 VHDL description of a 32 bit adder block

5.5.2 *Multiplier*

The multiplication of two binary numbers in fixed-point representation is performed in the same way as it is done for two decimal numbers. The difference is that multiplying two binary numbers involves logic 0's and 1's. On the other hand, as shown in Fig. 5.16, in a fixed-point representation one must include the bit of sign. In addition, one must be aware that multiplying two numbers (A, B) of N-bit will produce a result of $2N$-bit. Figure 5.17 shows the entity and the synthesis of the multiplier showing the input data with length of 32 bit and the output data is truncated to 32 bit also. This truncation operation is performed as already described in Chap. 2. This block also includes the pins of the CLK and RST to get a synchronous multiplier.

Listing 5.8 shows the VHDL description of the 32 bit multiplier with fixed-point format of 8.24. The multiplication operation is performed as follows: the 32 bit operands are loaded and the product is assigned to an internal signal of length equal to 64 bits ($r1$), after reset, and at the instant in which the ascendent flank of the CLK is produced, the resulting product is transferred to the output register but 32 bit are truncated and the other 32 bit are stored.

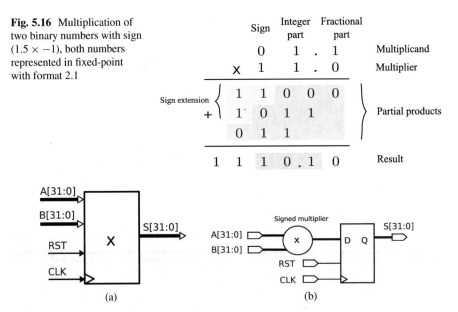

Fig. 5.16 Multiplication of two binary numbers with sign (1.5 × −1), both numbers represented in fixed-point with format 2.1

Fig. 5.17 (**a**) Entity of the multiplier block, and (**b**) architecture of the multiplier

```
1    library ieee;
2    use ieee.std_logic_1164.all;
3    use ieee.numeric_std.all;
4
5    entity Multiplier is
6         port(
7              CLK, RST : in std_logic;
8              A, B : in std_logic_vector(31 downto 0);
9              Output : out std_logic_vector(31 downto 0):=(others => '0'));
10   end Multiplier;
11
12   architecture behavioral of Multiplier is
13      signal r1: signed(63 downto 0);
14      begin
15         r1 <= signed(A)* signed(B);
16         process (CLK, RST) begin
17            if RST = '0' then
18               Output <= (others => '0');
19            elsif rising_edge(CLK)  then
20         Output <= std_logic_vector(R1(55 downto 24)); -- 32 bits
21            end if;
22         end process;
23   end behavioral;
```

Listing 5.8   VHDL description of a 32 bit Multiplier

5.5.3   Multiplexer

A multiplexer, commonly referred as MUX, is a widely used block in many digital
designs. Its main operation is selecting an output from a set of possible inputs,
and this selection depends on a control signal. Generally this block is used in the
digital implementation of chaotic oscillators to set the initial conditions and buffer
the values at each iteration of the associated numerical method.

Figure 5.18 shows the entity of a 6-inputs and 3-outputs multiplexer, also called
6:3 MUX. It includes a control signal labeled as Condition and the CLK and RST
pins. The multiplexer chooses between the six data inputs based on the condition as
shown in Table 5.8. Listing 5.9 shows the VHDL description of this multiplexer.

Fig. 5.18   Entity description
of a MUX including CLK and
RST pins

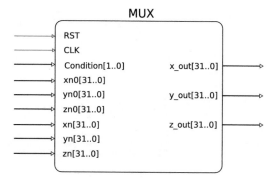

Table 5.8 Conditions controlling the multiplexer of Fig. 5.18

Condition		Output		
		x	y	z
0	0	others=> '0'	others=> '0'	others=> '0'
0	1	xn0	yn0	zn0
1	0	xn	yn	zn
1	1	others=> '0'	others=> '0'	others=> '0'

```vhdl
1   library ieee;
2   use ieee.std_logic_1164.all;
3
4   entity MUX is
5     port(
6     CLK, RST : in std_logic;
7     Condition: in std_logic_vector(1 downto 0);
8     xn0, yn0, zn0: in std_logic_vector(31 downto 0);
9     xn,  yn,  zn : in std_logic_vector(31 downto 0);
10    x, y, z: out std_logic_vector(31 downto 0):=(others=>'0'));
11  end MUX;
12
13  architecture Behavioral of MUX is
14
15  signal datox: std_logic_vector(31 downto 0);
16  signal datoy: std_logic_vector(31 downto 0);
17  signal datoz: std_logic_vector(31 downto 0);
18
19  begin
20     process(CLK,RST)
21        variable contador: integer:=0;
22         begin
23             if RST = '0' then
24                 datox <= (others=>'0');
25                 datoy <= (others=>'0');
26                 datoz <= (others=>'0');
27             elsif rising_edge(clk)  then
28         Case Condition is
29             when "00" =>
30                 datox<= (others=>'0');
31                 datoy<= (others=>'0');
32                 datoz<= (others=>'0');
33             when "01" =>
34                 datox<= xn0;
35                 datoy<= yn0;
36                 datoz<= zn0;
37     when "10" =>
38        datox<= xn;
39        datoy<= yn;
40        datoz<=zn;
41     when others =>
42        datox<= (others=>'0');
43        datoy<= (others=>'0');
44        datoz<= (others=>'0');
45         end Case;
46     end if;
47   end process;
48  x <= datox;
49  y <= datoy;
50  z <= datoz;
51  end Behavioral;
```

Listing 5.9 VHDL description of a 6:3 MUX

5.5.4   Counter

A sequential circuit is characterized by having memory elements, and its output depends on the combinations of the inputs, while it changes only after a change of the CLK pulse. An N-bit binary counter is a special class of sequential arithmetic with clock and reset inputs and an N-bit output. This kind of blocks is useful to access memory elements, count a number of CLK cycles to get an output of a function, etc. For instance, Fig. 5.19 shows the entity of a counter that can be programmed and whose function is controlled as shown in Table 5.9, where it can be observed that the counter has an input control with a pin called `Enable`, and depending on its logic value it can execute the actions listed therein: initializing the count to zero, counting, and hold the count. Listing 5.10 shows the VHDL description of a counter, in which when the count reaches `Count=255`, then the pin `Ready=1`.

```
1   library IEEE;
2   use IEEE.std_logic_1164.all;
3
4   entity Counter is
5     port(
6     RST, CLK: in  std_logic;
7     Enable  : in  std_logic_vector(1 downto 0);   --- Enable
8     Ready   : out std_logic;  --- Ready ='1' when the account ends
9     Count   : out std_logic_vector(8 downto 0));
10  end Counter;
11
12  Architecture Behavioral of Counter is
13  signal Q_1 : std_logic_vector(8 downto 0);
14  constant A : std_logic_vector(8 downto 0):="011111111";
15  begin
16      Count<=Q_1;
17      process(RST, CLK, Q_1)
18        begin
19            if RST='0' then
20                Q_1<=(others=>'0');
21      elsif rising_edge(CLK) then
22        case Enable is
23            when "00" =>
24                Q_1<=(others=>'0');
25            when "01" =>
26                Q_1<=Q_1+1;
27                if (Q_1 = 256 ) then
28                    Q_1 <=(others=>'0');
29                end if;
30            when "10" =>
31                Q_1<=Q_1;
32            when others =>
33          Q_1<=(others=>'0');
34            end case;
35
36        if Q_1=A then
37            Ready<='1';
38        else
39            Ready<='0';
40        end if;
41    end if;
42    end process;
43  end Behavioral;
```

Listing 5.10 VHDL description of an ascending counter

Fig. 5.19 Entity of a counter

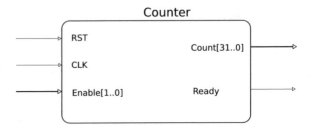

Table 5.9 Functions of a programmable counter

Enable		Action
0	0	Initializes the output to 0
0	1	Count
1	0	Hold
1	1	Initializes the output to 0

Fig. 5.20 Entity of a flip-flop type D

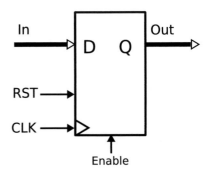

5.5.5 Flip-Flop

Flip-flops are widely used to store and transfer digital data. In particular the flip-flop type D is the most popular and most used to enhance the performance of a system by dividing large delays from combinational circuits and routing that are associated with critical paths. They execute an action in a single CLK cycle. The function of a flip-flop D is transferring the datum that comes from the input D to the output Q, and this is done after one CLK pulse. Figure 5.20 shows the entity of a flip-flop type D including an enable pin, and CLK and RST pins. Table 5.10 shows the truth table of the flanked flip-flop D, and including an `Enable=1`.

Table 5.10 Truth table of a
flip-flop type D including
enable pin

D	CLK	Enable	Q_n
0	↑	0	0
1	↑	0	0
x	x	0	0
0	↑	1	0
1	↑	1	1
x	x	1	Q_{n+1}

The VHDL description of the flip-flop type D in an array of blocks to process 32 bits is given in Listing 5.11.

```
1   library ieee;
2   use ieee.std_logic_1164.all;
3
4   entity FFD is
5       port (
6           CLK, RST: in std_logic;
7           D: in std_logic_vector(31 downto 0);
8           Q: out std_logic_vector(31 downto 0));
9   end FFD;
10
11  architecture Behavioral of FFD is begin
12      process (CLK,RST)
13      begin
14          if RST = '0' then
15              Q <= (others => '0');
16          elsif rising_edge(CLK) then
17          Q <= D;
18          end if;
19      end process;
20  end Behavioral;
```

Listing 5.11 Flip-flop D with ascendent flank

5.5.6   Cumulative Summation Block

The digital implementation of a summation like the one required to execute the approximation of Grünwald-Letnikov to simulate fractional-order chaotic oscillators is a challenge that involves the control of the memory length. It can be performed by the cumulative summation block shown in Fig. 5.21. It consists of one combinational multiplier, as the one described above, one combinational adder with enable, and two type-D registers to storage the cumulative sum. Let us consider the cumulative sum given in (5.13), which requires k consecutive multiplications and $k - 1$ addition operations per sample to compute the sum of products $S[k]$. Figure 5.22 shows the internal function of Fig. 5.21, from which after reset, and when clear = 1 and Px = 0, one can observe that $R_{3,j}$ is propagated through the summation. Once the $k - 1$ addition operations are done, the enable signal is set to Px = 1, and the data at D is assigned to the output $S[k]$.

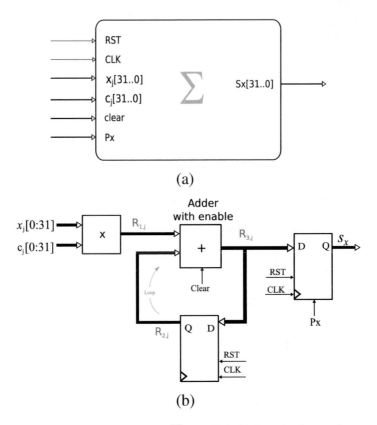

Fig. 5.21 (a) Entity of the cumulative sum ($\sum$), and (b) its hardware implementation

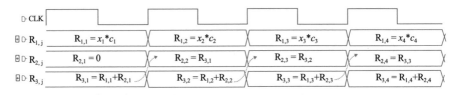

Fig. 5.22 Cumulative sum: timing diagram of its internal signals

$$S[k] = \sum_{j=v}^{k} c_j x_j \qquad (5.13)$$

The block shown in Fig. 5.21 results of great usefulness when the values of x_j and c_j are dynamic. It means that the result of $S[k]$ will change. For example in the case of the Grünwald-Letnikov method, it produces different values of x_j at each iteration, so that to ensure that any datum of the past iterations remains stored in the loop, one can design an adder with enable that initializes its output to 0. The

VHDL description of an adder with enable that can be used in Fig. 5.21 is given in Listing 5.12. This block has three inputs In1, In2, clear and one output Sx. If clear = 1, then its output is initialized to zero, otherwise the sum is executed with two's complement numbers in In1 and In2.

```vhdl
library ieee;
use ieee.std_logic_1164.all;
use ieee.std_logic_unsigned.all;

entity adder_clear is
   port(
       In1, In2 : in std_logic_vector(31 downto 0);
       clear : in std_logic;
       Sx : out std_logic_vector(31 downto 0):= (others => '0'));
end adder_clear;

architecture behavioral of adder_clear is
   signal suma: std_logic_vector(31 downto 0);
   begin
       Sx<=suma;
       process(clear, In1, In2)
          begin
             if (clear='1') then
                suma<=(others=>'0');
        else
                suma <= std_logic_vector(signed(In1) + signed(In2));
             end if;
          end process;
end behavioral;
```

Listing 5.12 VHDL description of a 32 bit adder with enable

5.6 Arrays for Designing Memories

Electronic memories come in many different formats and styles. The type of memory unit that is preferable for a given application is a function of the required memory size, the time it takes to access the stored data, the access patterns, and the system requirements [9].

5.6.1 Shift Register

Connecting flip-flops in an array is possible to create shift registers with the goal of storing any kind of information. This kind of structures are of great usefulness when requiring to storage temporal information. Figure 5.23 shows the architecture of a shift register, from which one can design a memory that allows to emulate the short memory principle.

Listing 5.13 shows the VHDL descriptions of four shift registers to generate a memory element. In this case, one uses the instruction instantiation of components (work.FFD(Behavioral)) that calls to the description of the flip-flop type D as

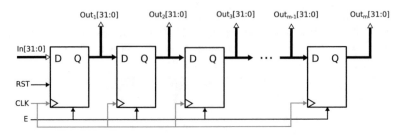

Fig. 5.23 Shift register designed with flip-flops

a subsystem. The shift operation is executed at each CLK pulse, and the data will be shifted from the `Input` to the right. The number of shifts will be equivalent to the number of stages or used registers. In this example four shifts are executed because there are four registers; however, the VHDL code can be easily adapted to increase the number of shifts.

```vhdl
library ieee;
use ieee.std_logic_1164.all;

entity shift_register is
    port (
        CLK, RST, E : in std_logic;
        Input: in std_logic_vector(31 downto 0);
        Output_1 : out std_logic_vector(31 downto 0);
        Output_2 : out std_logic_vector(31 downto 0);
        Output_3 : out std_logic_vector(31 downto 0);
        -- Add the rest of the outputs
        Output_256 : out std_logic_vector(31 downto 0));
end shift_register;

architecture behavioral of shift_register is

Signal R1: std_logic_vector(31 downto 0);
Signal R2: std_logic_vector(31 downto 0);
Signal R3: std_logic_vector(31 downto 0);
-- Add the rest of the signals
Signal R256: std_logic_vector(31 downto 0);

begin

    B1: entity work.FFD(Behavioral)
        port map (CLK, RST, E, Input, R1);
    B2: entity work.FFD(Behavioral)
        port map (CLK, RST, E, R1, R2);
    B3: entity work.FFD(Behavioral)
        port map (CLK, RST, E, R2, R3);
    -- Add the rest of the blocks
    B256: entity work.FFD(Behavioral)
        port map (CLK, RST, E, R3, R256);

    Output_1<=R1;
    Output_2<=R2;
    Output_3<=R3;
    -- Add the rest of assignments
    Output_256<=R256;

end Behavioral;
```

Listing 5.13 VHDL description of a shift register

5.6.2 Random Access Memories

The random access memories (RAMs) are volatile memories because one can read or write in a determined location of memory with the same wait time for any address. The majority of FPGAs integrate RAM modules, with different sizes, but in the majority the memory arrays will be specified by its memory depth × data width, as shown in Fig. 5.24a, where the depth of an array is the number of rows, and the width is the number of columns. For example, the Cyclone IV FPGA device family [10] has a RAM that includes memory blocks of type M9K, as shown in Fig. 5.24b, so that one can use the configuration of the width of the port like 256 × 32.

In Fig. 5.25 one can see the entity of a simple dual-port RAM, in which a dedicated address port is available for each read and write operation (`Addressr` and `Addressw`). In addition, this memory block contains ports to enable read (`rden`) and write (`wred`) operations. Listing 5.14 shows the VHDL description of this memory.

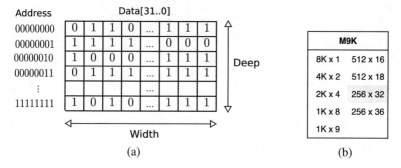

Fig. 5.24 RAM: (**a**) Memory size specification and (**b**) size configuration for Cyclone IV FPGA device family [10]

Fig. 5.25 Entity description of the RAM

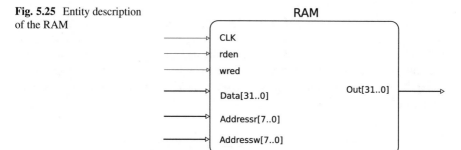

```
1   library ieee;
2   use ieee.std_logic_1164.all;
3   use ieee.std_logic_unsigned.all;
4
5   entity RAM_256 is
6       port (
7     clk : in std_logic;
8       rden : in std_logic;
9     wred : in std_logic;
10    data_in  : in std_logic_vector(31 downto 0);
11          addressw : in std_logic_vector(7 downto 0); --address to write
12          addressr : in std_logic_vector(7 downto 0); --address to read
13    data_out : out std_logic_vector(31 downto 0)
14      );
15  end RAM_256;
16
17  Architecture Behavioral of RAM_256 is
18      type memory_ram is array (0 to 255) of std_logic_vector (31 downto 0);
19      signal Memoryam : memory_ram:= (others => (others => '0'));
20      attribute ram_style: string;
21      attribute ram_style of Memoryam : signal is "M9K";
22      begin
23          memory : process (clk, rden, wred)
24              begin
25          if(rising_edge(clk)) then
26              if (wred = '1') then
27                  Memoryam(conv_integer(addressw)) <= data_in; -- write
28              end if;
29                  if (rden='1') then
30          data_out <= Memoryam(conv_integer(addressr)); -- read
31          else
32          data_out <=(others=>'0');
33              end if;
34      end if;
35          end process memory;
36  end behavioral;
```

Listing 5.14 VHDL description of a RAM

5.6.3 Read Only Memory

The read only memories (ROMs) allow only reading the stored data. These memories are efficient to storage data for a long time. Similar to the RAMs, the ROMs have different sizes usually specified by its memory depth × data width, and the number of ports. This chapter describes a single-port ROM for read operation. In Fig. 5.26 one can see the entity of a single-port ROM again of type M9K, so that the port is defined as 256 × 32.

Fig. 5.26 Entity of a single-port ROM

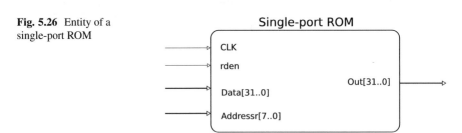

The VHDL description of the ROM based on M9K, with the port size 256×32, is given in Listing 5.15. It allows a storage of 256 data. When the signal port `rden=1` one can begin the reading of the information previously stored. The read operation uses read address `Addressr` and `Out` ports.

```vhdl
1   library ieee;
2   use ieee.std_logic_1164.all;
3   use ieee.std_logic_unsigned.all;
4
5   entity ROM_256 is
6      port (
7         clk, rden : in std_logic;
8         address_out : in std_logic_vector(7 downto 0);  --address to read
9         data_out : out std_logic_vector(31 downto 0));
10  end ROM_256;
11
12  Architecture Behavioral of ROM_256 is
13     type memory_rom is array (0 to 255) of std_logic_vector (31 downto 0);
14     signal Memoryom : memory_rom:=(
15        -- Insert 256 coefficients values for example:
16        -- x"FF11EB85",
17        -- x"FFF7AACE",
18        -- ...
19        );
20     begin
21        memory : process (clk, rden)
22        begin
23           if rising_edge(clk) then
24              if (rden='1') then
25                 data_out <= Memoryom(conv_integer(address_out));
26              else
27                 data_out <=(others=>'0');
28              end if;
29           end if;
30        end process memory;
31  end Behavioral;
```

Listing 5.15 VHDL description of one-port ROM

5.6.4   Lookup Table

A lookup table (LUT) is a fast way to realize a complex function in digital logic design. This block is very useful because it only requires to perform one search on the memory to represent any function. However, it requires a huge quantity of logic resources, especially when requiring high resolution of a function. On the other hand, as already shown in [11], it is unfortunately difficult to implement a good FPGA-based look-up table with more than 4 to 11 bit addresses. For instance, Fig. 5.27 shows the entity of a LUT having 8 bit address. Its VHDL description is given in Listing 5.16, which allows a storage of 7 data or 32 bits. This VHDL code can be easily adapted to storage mode values, like 255 or more.

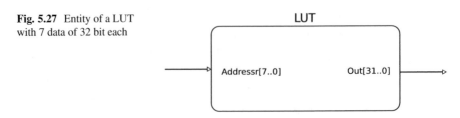

Fig. 5.27 Entity of a LUT with 7 data of 32 bit each

```
1   library ieee;
2   use ieee.std_logic_1164.all;
3
4   entity LUT is
5   port (
6       Address_01: in std_logic_vector(7 downto 0);
7       Output: out std_logic_vector(31 downto 0));
8   end LUT;
9
10  architecture behavioral of LUT is
11
12  signal Sout: std_logic_vector(31 downto 0);
13      begin
14        process(Address_01)
15          begin
16                case(Address_01) is
17                    when "00000000" => Sout <= x"F828F5C3";
18                    when "00000001" => Sout <= x"FFEBEDFA";
19                    when "00000010" => Sout <= x"FFF92D13";
20                    when "00000011" => Sout <= x"FFFC8DCD";
21                    when "00000100" => Sout <= x"FFFDEB27";
22                    when "00000101" => Sout <= x"FFFE9AFE";
23                    when others => Sout <= x"00000000";
24                end case;
25        end process;
26
27  Output <= Sout;
28  end Behavioral;
```

Listing 5.16 VHDL description of a LUT

5.7   FPGA-Based Implementation of Grünwald-Letnikov Method

The FPGA-based implementation of fractional-order chaotic oscillators requires an approximation method to simulate the fractional derivatives. These operations can be implemented using the VHDL descriptions given above, which can perform arithmetic and logic operations, as well as by using the memory blocks listed in Table 5.11. The implementation of Grünwald-Letnikov method can be improved exploiting the characteristics of these memories to reduce power consumption and area, and increase the speed of data processing. For example, shift registers can be used to save the values of $x(t_{k-j})$, $y(t_{k-j})$, $z(t_{k-j})$, and in this way implement

Table 5.11 Memory comparison

Memory type	Operation	Storage capacity	Latency	Transistors per bit cell
Flip-flop	Reading/writing	$x(t_{k-j})$, $y(t_{k-j})$, $z(t_{k-j})$	Fast	≈20 [7]
RAM	Reading/writing	$x(t_{k-j})$, $y(t_{k-j})$, $z(t_{k-j})$	Slow	6 per bit cell [12]
ROM	Reading	Binomial coefficients	Slow	≈4 per bit cell [13]
LUTs	Reading	Binomial coefficients	Slow	≈32 [14]

Table 5.12 Comparing FPGA resources when implementing Lorenz fractional-order chaotic oscillator using RAMs and shift registers

Resources	Available	RAM	Shift register
Total logic elements	146,760	2236	23,662
Total memory bits	6,635,520	24,576	135
Embedded multiplier 9 bit elements	720	92	720
Fmax (MHz)		80.87	87.13

a pseudo-random number generator with a latency equal to one CLK cycle, at expenses of increasing the logic resources and the power consumption.

One can also use LUTs, as shown in Table 5.11, to reach a latency of 1 CLK cycle as reported in [5]. However, the LUT complexity grows exponentially with the number of inputs, so that using a LUT with a large number of inputs as a logic block is unpractical [15], and it impacts in the area. Fortunately, many FPGA devices have been equipped with various specific digital blocks and memories, such as single/dual port RAMs and ROMs, multipliers, and digital signal processors (DSPs). These blocks are optimally designed and embedded in the devices to facilitate implementing specific functions that otherwise may require much larger number of LUTs to be implemented and provide an opportunity to implement applications with memory demand [6]. Therefore, one must take advantage of the characteristics of the RAM and ROM for the efficient processing of data. Another important point is that one must consider that the memory latency and throughput also depend on the memory size. That way, larger memories tend to be slower than smaller ones if all else is the same. Therefore, the best memory type for a particular design depends on the speed, cost, and power constraints [7].

5.7.1 FPGA-Based Implementation of Lorenz Fractional-Order Chaotic Oscillator Using RAMs and Shift Registers

The FPGA-based implementation of the Lorenz fractional-order chaotic oscillator can help to show the resources required for its design using RAMs and shift registers, as shown in Table 5.12. In this case, one can appreciate that a high number of memory elements trades the latency. For instance, considering using RAMs of the

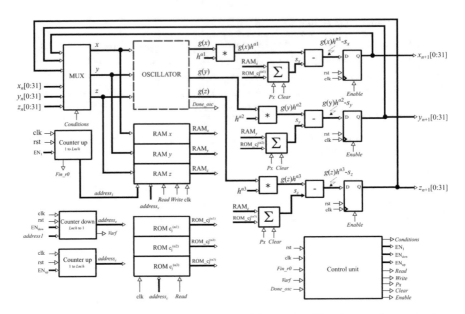

Fig. 5.28 Modular FPGA-based implementation of the Grünwald-Letnikov method to emulate any kind of fractional-order chaotic oscillator

type M9K with a port configuration by 256×32, then 256 CLK cycles will be needed to get the final value of the cumulative sum, and 258 CLK cycles are required to get the final results. In this case the latency is calculated with respect to the number of cycles that are necessary to get the final outputs of the state variables of the fractional-order chaotic oscillators, and the maximum frequency Fmax reached by the system. If the option is using shift registers, the latency will be significantly improved but the quantity of logic resources will increase, as well as the power consumption. That way, if the length of memory that is required is low to implement the Grünwald-Letnikov method, it is recommended to use shift registers, but if the required memory is large, then the best option is using the memory blocks already available in the FPGAs.

A proposed FPGA-based design is shown in Fig. 5.28. The block OSCILLATOR can be updated practically by any kind of fractional-order chaotic oscillator simulated by the Grünwald-Letnikov method. The short memory principle is implemented by using RAM and ROM blocks. With this proposed architecture, one can use different sizes of memories, basically modifying the Control unit and configuring the RAM and ROM blocks. In Fig. 5.28 the Grünwald-Letnikov method is implemented in a modular way to change the different fractional-order chaotic oscillators along the associated ROM. The Control unit is a finite state machine (FSM) that controls the counters, read or write mode in the memory blocks, the multipliers, and the selection for the multiplexors. The general operation of this control unit is presented in Algorithm 1, and its operation can be described as follows:

1. The maximum memory length is determined and the coefficient values are precharged in a ROM.
2. All inputs and outputs for each entity are initialized to zero. Also all control variables are initialized to zero.
3. While input *rst* is different than a logic-"0," the system remains in a waiting state. With *rst* in logic-"1" its action is to validate if it is the first time the system is executed, using a multiplexer with the control signal called *Conditions*.

   With *Conditions* = 0 the initial values for x_n, y_n, z_n are charged, and then the control unit assigns a logic-"1" to *Conditions*. When *Conditions* = 1, the outputs $x_{n+1}, y_{n+1}, z_{n+1}$ that will feedback to the system are charged. The outputs of the multiplexer are set as x, y, z.
4. The address *address$_l$* of the RAM is set using a counter. After this, writing is enabled with RAM's *write* = 1 in which x, y, z values will be stored. Once the write ends, its signal is disabled with *write* = 0.
5. The values of x, y, z will be evaluated and their result $f(x), f(y), f(z)$ will be multiplied by $h^{\alpha_1}, h^{\alpha_2}, h^{\alpha_3}$. The result will be stored in variable *eval*.
6. Concurrently, the cumulative sum shown in Fig. 5.21 is activated. This block calculates the multiplication of the coefficients with variables x, y, z that are in both ROM and RAM. To perform this operation, the address for read on memory is set, and the addresses for the coefficients *address$_c$* starts in one, and each CLK cycle is incremented until the size *memory* is filled. The initial value of variables' addresses, *address$_v$*, is always the last address *address$_l$* where the write operation was performed and in each CLK cycle its value is decremented inmodulus *memory*.
7. While *address$_c$* ≤ *memory* is in reading mode it is activated. The cumulative sum with all the values stored in the memory blocks is performed.
8. When *address$_c$* > *memory*, the values of the cumulative sum and *eval* are subtracted.

5.7.2 Experimental Observation of Fractional-Order Chaotic Attractors

This subsection shows the experimental observation of the fractional-order chaotic attractors of the oscillators listed in Table 5.2. Figure 5.29 shows the experimental results of the phase-space portraits of the FPGA-based implementation of the fractional-order Lorenz chaotic oscillator; Fig. 5.30 shows the phase-space portraits of the fractional-order chaotic attractors of Rössler oscillator; Fig. 5.31 shows the phase-space portraits of the fractional-order chaotic attractors of Chen oscillator; and Fig. 5.32 shows the phase-space portraits of the fractional-order chaotic attractors of Lü oscillator. All these experimental fractional-order chaotic attractors were implemented applying the Grünwald-Letnikov numerical method, and Table 5.13 shows different chaotic responses of Lorenz fractional-order chaotic oscillator using

Algorithm 1 Controlling the processes of the Grünwald-Letnikov method for the FPGA-based implementation of fractional-order chaotic oscillators

---

$memory \leftarrow memory\ size$

2: *Initialize to zero*

   while RST=1 **do**

4:    **if** *Conditions* **then**

       $x, y, z \leftarrow x_{n+1}, y_{n+1}, z_{n+1}$

6:    **else**

       $x, y, z \leftarrow x_n, y_n, z_n$

8:       $Conditions = 1$

    end if

10:   $address_l \leftarrow address_l + 1$ *{**Address for** x, y, z}*

     $write = 1$

12:   $RAM(address_v, write) \leftarrow x, y, z$

     $write = 0$

14:   $[f(x), f(y), f(z)] \leftarrow oscillator\ (x, y, z)$

     $eval \leftarrow [h^{q1} f(x), h^{q2} f(y), h^{q3} f(z)]$

16:   $Clear = 1$

     $address_v \leftarrow address_l$

18:   $address_c \leftarrow 1$ *{**Address for** $Cj^{(q1)}, Cj^{(q2)}, Cj^{(q3)}$}*

    while $address_c \leq memory$ **do++**

20:      $read = 1$

       $(x, y, z) \leftarrow RAM(address_v, read)$

22:      $[C_j^{(q1)}, C_j^{(q2)}, C_j^{(q3)}] \leftarrow ROM(address_c, read)$

       $(s_x, s_y, s_z) \leftarrow cumulative\_sum(x, y, z, C_j^{(q1)}, C_j^{(q2)}, C_j^{(q3)})$

24:      $address_v \leftarrow address_v - 1$

       $address_c \leftarrow address_c + 1$

26:    **end while**

     $Clear = 0$

28:   $read = 0$

     $(x_{n+1}, y_{n+1}, z_{n+1}) \leftarrow eval - (s_x, s_y, s_z)$

30:   $Output \leftarrow (x_{n+1}, y_{n+1}, z_{n+1})$

    end while

FPGA architectures with different memory length (L) and step-sizes (h). As one can see, the maximum value of the memory length L is 256, and the whole available RAM in the FPGA-based implementation is 256×32. Again, as it was the case for the numerical simulation discussed above, the experimental implementation on FPGA shows the same dynamical behavior: the dynamics of the Lorenz fractional-order system decrease as h decreases under a fixed L, and when L is small and as h decreases, then the chaotic behavior tends to disappear. This trade-off between guaranteeing chaotic dynamics and reducing hardware resources is a challenge to estimate it from an analytical equation.

    The experimental results to observe the fractional-order chaotic attractors were obtained using the Altera Cyclone IV GX FPGA DE2i-150, and a 16 bit DAC DAS1612VA, which has a speed of 200 kSamples/s. Considering that the FPGA-based implementation of the fractional-order chaotic oscillators is done using

Fig. 5.29 Phase-space portraits of the FPGA-based implementation of the fractional-order Lorenz chaotic oscillator applying Grünwald-Letnikov method. Experimental attractors in the: (**a**) X–Y view, (**b**) X–Z view, and (**c**) Y–Z view with axes 1 V–1 V/division in a Lecroy oscilloscope

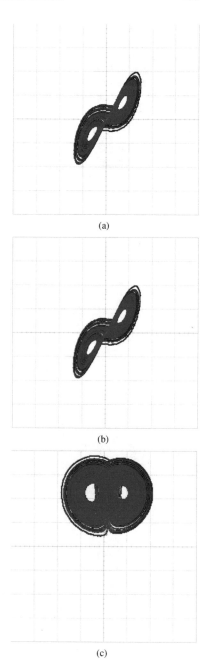

(a)

(b)

(c)

Fig. 5.30 Phase-space portraits of the FPGA-based implementation of the fractional-order Rössler chaotic oscillator applying Grünwald-Letnikov method. Experimental attractors in the: (**a**) X–Y view with 500 mV–1 V/division, (**b**) X–Z view, and (**c**) Y–Z view with axes 500 mV–2 V/division in the oscilloscope

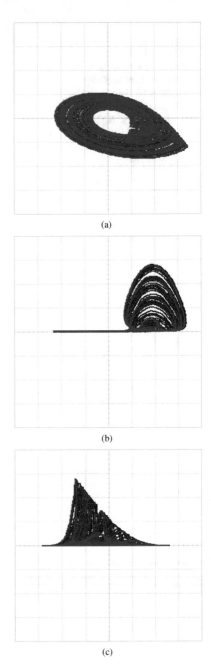

(a)

(b)

(c)

Fig. 5.31 Phase-space portraits of the FPGA-based implementation of the fractional-order Chen chaotic oscillator applying Grünwald-Letnikov method. Experimental attractors in the: (**a**) X–Y view, (**b**) X–Z view, and (**c**) Y–Z view with axes 1 V–1 V/division in the oscilloscope

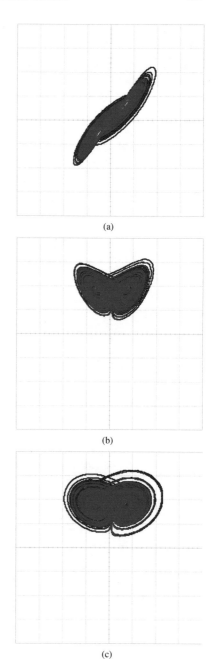

(a)

(b)

(c)

Fig. 5.32 Phase-space portraits of the FPGA-based implementation of the fractional-order Lü chaotic oscillator applying Grünwald-Letnikov method. Experimental attractors in the: (**a**) X–Y view, (**b**) X–Z view, and (**c**) Y–Z view with axes 1 V–2 V/division in the oscilloscope

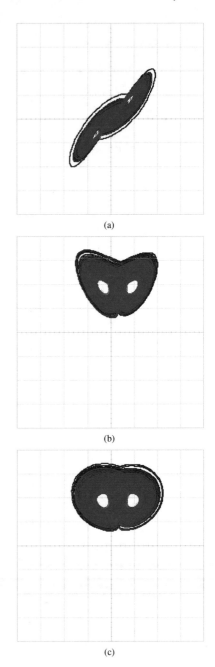

(a)

(b)

(c)

Table 5.13 Experimental observation of the fractional-order chaotic attractors of Lorenz system applying Grünwald-Letnikov method with different memory length L and step-sizes h

computer arithmetic of 32 bits, the DAC has a truncation of 16 bits for the state variables $(x_{n+1} \; y_{n+1} \; z_{n+1})$, as already described in Chap. 2. One must recall that the truncation must be performed considering the location of the point in the fixed-point notation; otherwise one cannot observe the experimental attractor, as one can see in Fig. 2.22.

5.8   FPGA-Based Implementation of Adams-Bashforth-Moulton Method

A more elaborated approximation to simulate fractional-order chaotic oscillators is the well-known Adams-Bashforth-Moulton method, which is a predictor-corrector type, similar to the multistep methods discussed in Chap. 1. The authors in [1] described this predictor-corrector approximation, mentioning its relation to Caputo's derivative, because it just requires the initial conditions and for an unknown function it has clear physical meaning. The Adams-Bashforth-Moulton method is based on the fact that the fractional differential equation given in (5.14) is equivalent to the Volterra integral equation given in (5.15), where it can be appreciated that the sum outside the integral, on the right-hand side, is completely determined by the initial values and hence is known.

$$D_t^\alpha y(t) = f(y(t), t), \; y^{(k)}(0) = y_0^{(k)}, \quad k = 0, 1, \ldots, m - 1 \tag{5.14}$$

$$y(t) = \sum_{k=0}^{[\alpha]-1} y_0^{(k)} \frac{t^k}{k!} + \frac{1}{\Gamma(\alpha)} \int_0^t (t - \tau)^{\alpha-1} f(\tau, y(\tau)) d\tau \tag{5.15}$$

Discretizing (5.15) for uniform grid $t_n = nh$, with $n = 0, 1, \ldots, N$ and $h = Tsim/N$, and using the short memory principle, one can obtain a close numerical approximation of the true solution $y(t_n)$ of the fractional differential equation (5.14), while preserving the order of accuracy. Let us assume that one have calculated the approximation $y_h(t_j)$, $j = 1, 2, \ldots, n$, and one wants to obtain $y_h(t_{n+1})$ by means of the equation given in (5.16), where $a_{j,n+1}$ is evaluated by (5.17), then the preliminary approximation to $y_h^P(t_{n+1})$ is called the predictor and is given by (5.18), where $b_{j,n+1}$ is evaluated by (5.19).

$$y_h(t_{n+1}) = \sum_{k=0}^{m-1} \frac{t_{n+1}^k}{k!} y_0^{(k)} + \frac{h^\alpha}{\Gamma(\alpha + 2)} f(t_{n+1}, y_h^P(t_{n+1}))$$

$$+ \frac{h^\alpha}{\Gamma(\alpha + 2)} \sum_{j=0}^{n} a_{j,n+1} f(t_j, y_n(t_j)) \tag{5.16}$$

$$a_{j,n+1} = \begin{cases} n^{\alpha+1} - (n-\alpha)(n+1)^\alpha, & \text{if } j = 0 \\ (n-j+2)^{\alpha+1} + (n-j)^{\alpha+1} + 2(n-j+1)^{\alpha+1}, & \text{if } 1 \le j \le n \\ 1, & \text{if } j = n+1 \end{cases}$$

$$(5.17)$$

$$y_h^P(t_{n+1}) = \sum_{k=0}^{m-1} \frac{t_{n+1}^k}{k!} y_0^{(k)} + \frac{1}{\Gamma(\alpha)} \sum_{j=0}^{n} b_{j,n+1} f(t_j, y_n(t_j)) \tag{5.18}$$

$$b_{j,n+1} = \frac{h^\alpha}{\alpha} ((n+1-j)^\alpha - (n-j)^\alpha) \tag{5.19}$$

The equation given in (5.16) can be applied to approximate the numerical solution of a fractional-order chaotic system. For example, let us consider the following fractional-order system given in (5.20) [16], with $0 < \alpha_i \le 1$ for $i = 1, 2, 3$, and initial condition $(x_0\ y_0,\ z_0)$. According to this approximation method, the fractional-order system given in (5.20) can be discretized as shown in (5.21), where the predictors must evaluate the equations given in (5.22).

$$D^{\alpha_1} x(t) = f_1(x, y, z)$$
$$D^{\alpha_2} y(t) = f_2(x, y, z) \tag{5.20}$$
$$D^{\alpha_3} z(t) = f_3(x, y, z)$$

$$x_{n+1} = x_0 + \frac{h^{\alpha_1}}{\Gamma(\alpha_1 + 2)} f_1(x_{n+1}^P, y_{n+1}^P, z_{n+1}^P)$$

$$+ \frac{h^{\alpha_1}}{\Gamma(\alpha_1 + 2)} \sum_{j=0}^{n} a_{j,n+1} f_1(x_j, y_j, z_j)$$

$$y_{n+1} = x_0 + \frac{h^{\alpha_2}}{\Gamma(\alpha_2 + 2)} f_2(x_{n+1}^P, y_{n+1}^P, z_{n+1}^P)$$

$$+ \frac{h^{\alpha_2}}{\Gamma(\alpha_2 + 2)} \sum_{j=0}^{n} a_{j,n+1} f_2(x_j, y_j, z_j) \tag{5.21}$$

$$z_{n+1} = x_0 + \frac{h^{\alpha_3}}{\Gamma(\alpha_3 + 2)} f_3(x_{n+1}^P, y_{n+1}^P, z_{n+1}^P)$$

$$+ \frac{h^{\alpha_3}}{\Gamma(\alpha_3 + 2)} \sum_{j=0}^{n} a_{j,n+1} f_3(x_j, y_j, z_j)$$

$$x_{n+1}^p = x_0 + \frac{1}{\Gamma(\alpha_1)} \sum_{j=0}^{n} b_{j,n+1} f_1(x_j, y_j, z_j)$$

$$y_{n+1}^p = x_0 + \frac{1}{\Gamma(\alpha_2)} \sum_{j=0}^{n} b_{j,n+1} f_2(x_j, y_j, z_j) \qquad (5.22)$$

$$z_{n+1}^p = x_0 + \frac{1}{\Gamma(\alpha_3)} \sum_{j=0}^{n} b_{j,n+1} f_3(x_j, y_j, z_j)$$

Henceforth, from the Adams-Bashforth-Moulton method that approximates a fractional-order system applying (5.21) and (5.22), with $a_{j,n+1}$ and $b_{j,n+1}$ evaluated by (5.17) and (5.19), respectively, then one can implement fractional-order chaotic oscillators on an FPGA.

5.8.1   Simulating Fractional-Order Chaotic Oscillators Applying Adams-Bashforth-Moulton Approximation

The application of the Adams-Bashforth-Moulton method using (5.16) helps to simulate fractional-order chaotic oscillators, as the ones listed in Table 5.2, and already simulated applying Grünwald-Letnikov method. Recall that the Adams-Bashforth-Moulton method requires computing values by a predictor approximation and then the final values at the actual iteration is evaluated by (5.16) that is called corrector. This is the reason why this approximation is of predictor-corrector type.

The numerical simulations programming the Adams-Bashforth-Moulton method using (5.16) are shown in the following figures. For example, Fig. 5.33 shows the phase-space portraits among two state variables for the fractional-order Lorenz chaotic oscillator, Fig. 5.34 shows the simulated attractors for the fractional-order Rössler chaotic oscillator, Fig. 5.35 shows the results of simulating the fractional-order Chen chaotic oscillator, Fig. 5.36 shows the fractional-order chaotic attractors for the Lü system, and Fig. 5.37 shows the simulation performed for the fractional-order V-Shape chaotic oscillator. In all cases the simulation conditions were taken from Table 5.2.

5.8.2   Effects of the Short Memory Principle in Adams-Bashforth-Moulton Method

As discussed before, the solution of a fractional-order chaotic oscillator requires a huge size of memory to guarantee the best exactness to the real solution. However, in real applications the memory is finite so that one must deal with

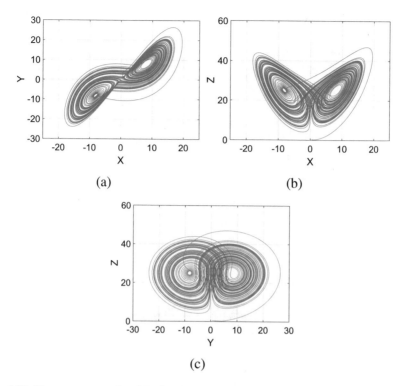

Fig. 5.33 Phase-space portraits of the fractional-order Lorenz chaotic system: (**a**) X–Y view, (**b**) X–Z view, and (**c**) Y–Z view

other approximations to reduce hardware resources, and in this case one can apply the short memory principle already discussed before. This fact allows a hardware designer to approximate the numerical solution by using the information of the "recent past." It means that the interval $[t - L, t]$ can be associated with a finite length of memory L. For example, applying the short memory principle requires the approximation of the fractional-derivative as given by (5.23) [4], where the number of terms to add is limited by the value of L/h. In this case, the error of the approximation when $|f(t)| \ll M, (0 < t \ll t_1)$ is bounded and expressed by (5.24), which can be used to determine the necessary memory length to obtain a certain error bound. Compared to the short memory principle applied to the Grünwald-Letnikov method as shown in Table 5.13, the Adams-Bashforth-Moulton method requires a higher length of memory L, because as t increases, the "recent past information" has a higher influence on the solutions, as shown in the following subsections.

$$D^\alpha f(t) \approx_{t-L} D^\alpha f(t), \quad t > L \tag{5.23}$$

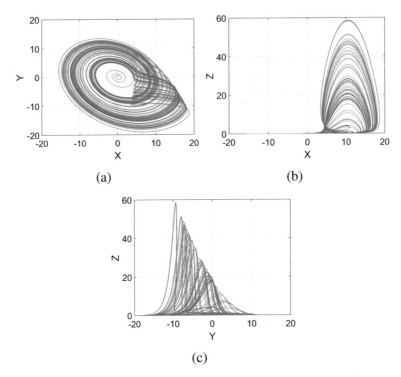

Fig. 5.34 Phase-space portraits of the fractional-order Rössler chaotic system: (**a**) X–Y view, (**b**) X–Z view, and (**c**) Y–Z view

$$\epsilon(t) = |D^\alpha f(t) -_{t-L} D^\alpha f(t)| \ll \left(\frac{ML^{-\alpha}}{|\Gamma(1-\alpha)|} \right)^{1/\alpha} \tag{5.24}$$

Analyzing (5.17) and (5.19), it can be observed that $a_{j,n+1}$ and $b_{j,n+1}$ will have dynamical values as j reaches n ($j = 0 \rightarrow n$). For example, let us consider the fractional-order Chen chaotic oscillator given in Table 5.2, which is simulated using $a = 35, b = 3, c = 28$, and initial conditions $(-9, -5, 14)$ for a commensurate fractional-order system $(0.9, 0.9, 0.9)$ and a step-size of $h = 0.002$. Discretizing (5.21) for uniform grid $t_n = nh$ with different values of n, then the coefficients $b_{j,n+1}$ evolve as shown in Fig. 5.38. In a similar way, the coefficients $a_{j,n+1}$ evolve as shown in Fig. 5.39. Recall that the binomial coefficients in the Grünwald-Letnikov method tend to 0 but in Adams-Bashforth-Moulton they have unstable behavior as n increases, so that one should use a higher length of memory.

The effect of using different lengths of memory is shown in Table 5.14 for the simulation of the fractional-order Chen chaotic oscillator using the conditions mentioned above. It can be appreciated that the required memory L is much higher than when applying the Grünwald-Letnikov method, and the dynamics of the fractional-order chaotic attractor is much better as n increases.

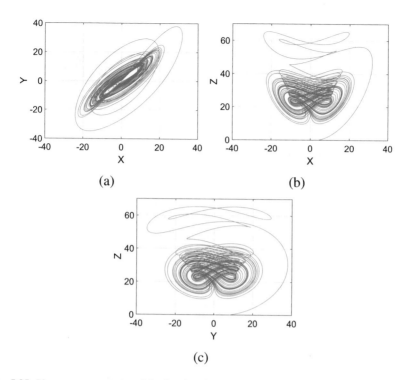

Fig. 5.35 Phase-space portraits of the fractional-order Chen chaotic system: (**a**) X–Y view, (**b**) X–Z view, and (**c**) Y–Z view

Table 5.14 Simulation of the fractional-order Chen chaotic oscillator with Adams-Bashforth-Moulton using different lengths of memory

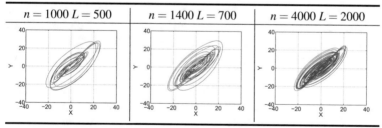

$n = 1000\ L = 500$	$n = 1400\ L = 700$	$n = 4000\ L = 2000$

5.8.3   On the FPGA-Based Implementation of Adams-Bashforth-Moulton Method

The VHDL description of the Adams-Bashforth-Moulton method can be performed using the entities and architectures detailed above for the Grünwald-Letnikov method. The implementation in this case can be done in a modular architecture, so that one can change just the kind of fractional-order chaotic oscillator and the

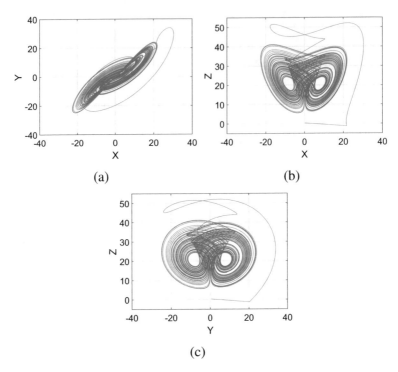

Fig. 5.36 Phase-space portraits of the fractional-order Lü chaotic system: (**a**) X–Y view, (**b**) X–Z view, and (**c**) Y–Z view

length of memory to generate fractional-order chaotic attractors. The numerical representation is also in fixed-point format of 32 bit and in two's complement. The VHDL description begins from the observation of the discretization of the fractional-order chaotic oscillator, for example from (5.21). The first step should be determining the values of $f_1(x_j,\ y_j,\ z_j)$, $f_2(x_j,\ y_j,\ z_j)$, and $f_3(x_j,\ y_j,\ z_j)$, which can be associated with the evaluation of $f(x_j)$, $f(y_j)$, and $f(z_j)$. These values can be associated to an entity that is called Oscillator, in which all the equations describing the fractional-order chaotic oscillator are implemented. Let us consider again the fractional-order Chen chaotic oscillator, which description as a block of type Oscillator should be as the one shown in Fig. 5.40.

As described before, the Adams-Bashforth-Moulton method requires evaluating two stages, one associated with the predictor approximations and the other to the corrector one. For the former case, Fig. 5.41 shows the synthesis of the predictor entity that provides the values of the state variables labeled as x^p_{n+1}, y^p_{n+1}, z^p_{n+1}, which are given in (5.22). The evaluation of these state variables for the block associated with the predictor stage needs the use of ROMs to store the values of the coefficients $b_{j,n+1}$, and also it requires the use of RAMs to store the outputs of the architecture that is called Oscillator. Those outputs have dynamical

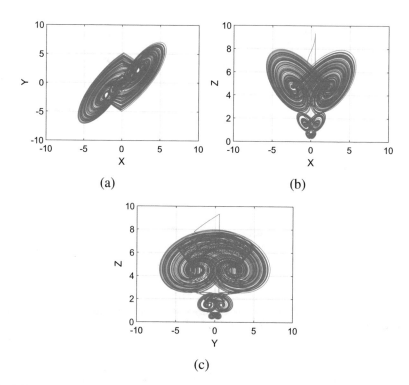

Fig. 5.37 Phase-space portraits of the fractional-order V-Shape chaotic system: (**a**) X–Y view, (**b**) X–Z view, and (**c**) Y–Z view

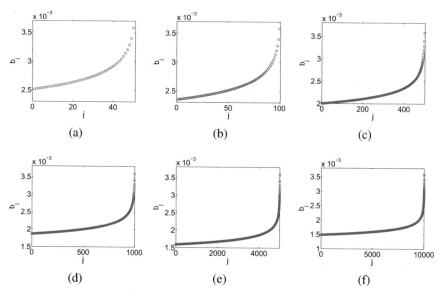

Fig. 5.38 Evolutions of coefficients $b_{j,n+1}$ for: (**a**) $n = 50$, (**b**) $n = 100$, (**c**) $n = 500$, (**d**) $n = 1000$, (**e**) $n = 5000$, and (**f**) $n = 10,000$

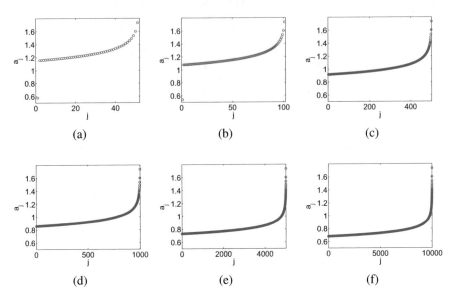

Fig. 5.39 Evolution of coefficients $a_{j,n+1}$ for: (**a**) $n = 50$, (**b**) $n = 100$, (**c**) $n = 500$, (**d**) $n = 1000$, (**e**) $n = 5000$, and (**f**) $n = 10{,}000$

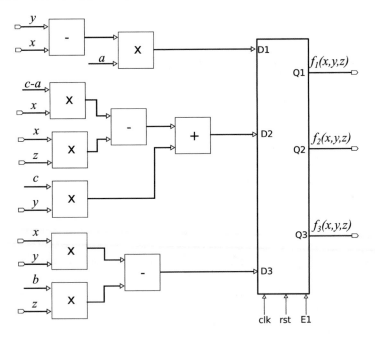

Fig. 5.40 Block description of the entity Oscillator for the fractional-order Chen chaotic oscillator

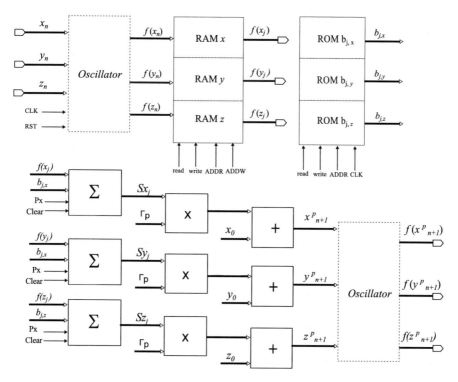

Fig. 5.41 Synthesis of the predictor stage in the Adams-Bashforth-Moulton method

values and are used in the second stage of the Adams-Bashforth-Moulton method, in the corrector stage, which is associated with the evaluation of (5.22) and (5.21). The sum operation is performed using the cumulative sum block described in the previous VHDL description of the Grünwald-Letnikov method. When the read process in the RAM and ROM is finished, the final value of the cumulative sum block is multiplied by $\Gamma_p = \frac{h^{\alpha_1}}{\Gamma(\alpha_1)}$, and to the product result one should add the value of the initial condition. The `entity` of the synthesized hardware for the predictor stage of the Adams-Bashforth-Moulton method is shown in Fig. 5.41, and it can be encapsulated into a high-level abstraction, so that the block describing the input and output ports is shown in Fig. 5.42. This `entity` labeled `Oscillator` for the predictor stage is used in the whole architecture that is shown in Fig. 5.43.

As mentioned above, the entity `Oscillator` is used to predict the values of $f(x_{n+1}^P)$, $f(y_{n+1}^P)$, $f(z_{n+1}^P)$, and $f(x_j)$, $f(y_j)$, $f(z_j)$. In Fig. 5.43 it can be noted that ROMs are used to store the values of the coefficients $a_{j,x}$ for each state variable, and again, the cumulative sum block is used to perform another summation but for the corrector stage, and as mentioned before, the result of this evaluation process is multiplied by $\Gamma_c = \frac{h^{\alpha_1}}{\Gamma(\alpha_1+2)}$. The values $f(x_{n+1}^P)$, $f(y_{n+1}^P)$, $f(z_{n+1}^P)$ are also multiplied by Γ_c. Finally, both results are added and afterwards, the result is added to x_0. It is important mentioning that the number of clock cycles that are required to

Fig. 5.42 Entity description of the architecture of the predictor block shown in Fig. 5.41

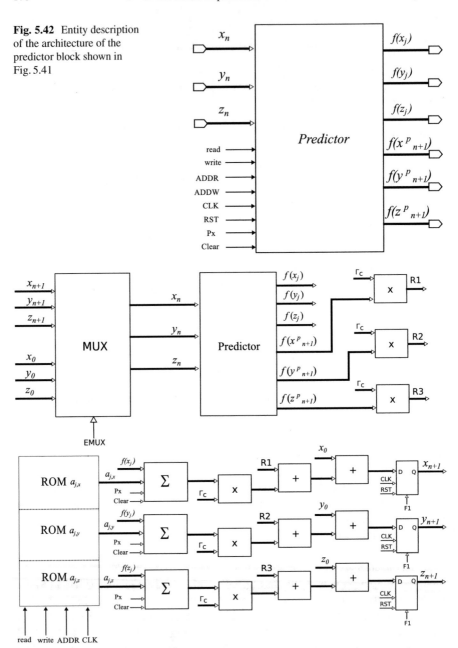

Fig. 5.43 Synthesis of the Adams-Bashforth-Moulton method that is based on the predictor-corrector operations. The prediction is performed by the entity associated with the block predictor, already described in Figs. 5.41 and 5.42

generate the next value for the whole iteration will depend on the type of fractional-order chaotic oscillator. In this case, the control of the iterations is performed by a finite state machine (FSM) that controls the counters that set the address to the RAM and ROM. The FSM is also designed to control the arithmetic operations of the multipliers, the register, and multiplexer. In short, the general control operations in the FPGA-based implementation of the Adams-Bashforth-Moulton method can be summarized as follows.

1. The length of the memory is estimated to describe in VHDL its maximum size. In this case for the Adams-Bashforth-Moulton method the depth is declared as `memory=256`.
2. All the inputs and outputs of each entity or digital block are initialized to zero, including the control variables.
3. While the inputs `RST` in Fig. 5.43 are different to logic "0," the FPGA-based architecture will be standby, and when `RST` changes to logic "1" the Adams-Bashforth-Moulton method will begin to generate the fractional-order chaotic attractor. In the FPGA-based implementation of the fractional-order chaotic oscillator, a multiplexer is used to control the signals that are processed at the input of the architecture. For instance, when `EMUX="01"` the initial conditions of the fractional-order chaotic oscillator (x_0, y_0, z_0) are charged. When `EMUX="10"` the outputs of the FPGA-based architecture (x_{n+1}, y_{n+1}, z_{n+1}) are feedback to the input ports. The outputs of the multiplexer block are established as (x_n, y_n, z_n).
4. The values of the state variables (x_n, y_n, z_n) are transferred to the `entity Oscillator` that corresponds to the predictor stage to evaluate $f(x_n)$, $f(y_n)$ $f(z_n)$.
5. The address `ADDW` in the RAM is configured using a counter. Afterwards, the write signal is enabled `write='1'` so that the values of $f(x_n)$, $f(y_n)$, $f(z_n)$ are stored. Finishing the written the signal changes to `write='0'`. The outputs of the RAMs are labeled as $f(x_j)$, $f(y_j)$, $f(z_j)$.
6. The blocks corresponding to the cumulative sum Σ in Figs. 5.41 and 5.43 are activated in order to perform the multiplication of the coefficients $a_{j,x}f(x_j)$, $a_{j,y}f(y_j)$, $a_{j,z}f(z_j)$ and $b_{j,x}f(x_j)$, $b_{j,y}f(y_j)$, $b_{j,z}f(z_j)$. Besides, the coefficients $a_{j,\{x,y,z\}}$ and $b_{j,\{x,y,z\}}$ are previously stored in the ROMs. To do this operation, the read address is set by `ADDR`, in which the address of the ROMs begin with zero and it increases at each clock cycle until reaching the previously established memory size. In the case of the address of the RAMs, the read operation is performed in each clock cycle and its value is settled by a descendent counter.
7. While `ADDR<=memory` the lecture mode will remain active and the cumulative sum will be executed with all the stored values within the memory blocks.
8. When `ADDR>memory` the FPGA-based architecture will provide the final result of the cumulative sum, and it will be multiplied by Γp and Γc, in both stages predictor and corrector in the Adams-Bashforth-Moulton method.

9. In the predictor stage, the result of the multiplications $\Gamma p S x_j$, $\Gamma p S y_j$, $\Gamma p S z_j$ will be added to the values of the initial conditions to get x_{n+1}^p, y_{n+1}^p, z_{n+1}^p. These predicted values of the state variables will be used to evaluate the functions in the `Oscillator` entity. In this particular case, the outputs of that block are then labeled as $f(x_{n+1}^p)$, $f(y_{n+1}^p)$, $f(z_{n+1}^p)$.

10. The values of $f(x_{n+1}^p)$, $f(y_{n+1}^p)$, $f(z_{n+1}^p)$ are multiplied by Γc, thus obtaining `R1`, `R2`, `R3`.

11. The values of `R1`, `R2`, `R3` are added to the results that are provided by performing the operations $\Gamma c f(x_j) a_{j,x}$, $\Gamma c f(y_j) a_{j,y}$, $\Gamma c f(z_j) a_{j,z}$. Finally, this result is added to the initial conditions and the signal `F1` is activated.

One must recall that the FPGA-based implementation of the Adams-Bashforth-Moulton method requires establishing a fixed number for n, so that the lengths of the memories sizes must be updated according to the type of fractional-order chaotic oscillator. For the case of the fractional-order Chen chaotic oscillator, the memory length is equal to 256, which requires 272 clock cycles to obtain the values of $(x_{n+1}, y_{n+1}, z_{n+1})$. In that case, 8 clock cycles correspond to the entity `Oscillator`. However, this number of clock cycles will change depending on the type of fractional-order chaotic oscillator.

References

1. W. Deng, Short memory principle and a predictor–corrector approach for fractional differential equations. J. Comput. Appl. Math. **206**(1), 174–188 (2007)
2. I. Petráš, *Fractional-Order Nonlinear Systems: Modeling, Analysis and Simulation* (Springer, Berlin, 2011)
3. D.K. Shah, R.B. Chaurasiya, V.A. Vyawahare, K. Pichhode, M.D. Patil, FPGA implementation of fractional-order chaotic systems. AEU Int. J. Electron. Commun. **78**, 245–257 (2017)
4. C.A. Monje, Y. Chen, B.M. Vinagre, D. Xue, V. Feliu-Batlle, *Fractional-Order Systems and Controls: Fundamentals and Applications* (Springer, Berlin, 2010)
5. M.F. Tolba, A.M. AbdelAty, N.S. Soliman, L.A. Said, A.H. Madian, A.T. Azar, A.G. Radwan, FPGA implementation of two fractional order chaotic systems. AEU Int. J. Electron. Commun. **78**, 162–172 (2017)
6. Z. Seifoori, Z. Ebrahimi, B. Khaleghi, H. Asadi, Introduction to emerging SRAM-based FPGA architectures in dark silicon era, in *Advances in Computers*, vol. 110 (Elsevier, San Diego, 2018), pp. 259–294
7. D. Harris, S. Harris, *Digital Design and Computer Architecture* (Morgan Kaufmann, Amsterdam, 2010)
8. W.J. MacLean, An evaluation of the suitability of FPGAs for embedded vision systems, in *2005 IEEE Computer Society Conference on Computer Vision and Pattern Recognition (CVPR'05)-Workshops* (IEEE, Piscataway, 2005), pp. 131–131
9. J.M. Rabaey, A.P. Chandrakasan, B. Nikolic, *Digital Integrated Circuits*, vol. 2 (Prentice Hall, Englewood Cliffs, 2002)
10. Internal Memory (RAM and ROM) user guide. https://www.intel.com/content/dam/www/programmable/us/en/pdfs/literature/an/an207.pdf
11. U. Meyer-Baese, *Digital Signal Processing with Field Programmable Gate Arrays*, vol. 65 (Springer, Berlin, 2007)

12. X. Dong, X. Wu, G. Sun, Y. Xie, H. Li, Y. Chen, Circuit and microarchitecture evaluation of 3D stacking magnetic RAM (MRAM) as a universal memory replacement, in *2008 45th ACM/IEEE Design Automation Conference* (IEEE, Piscataway, 2008), pp. 554–559
13. D. Etiemble, M. Israel, Comparison of binary and multivalued ICs according to VLSI criteria. Computer **21**(4), 28–42 (1988)
14. A. Sheikholeslami, R. Yoshimura, P.G. Gulak, Look-up tables (LUTs) for multiple-valued, combinational logic, in *Proceedings. 1998 28th IEEE International Symposium on Multiple-Valued Logic (Cat. No. 98CB36138)* (IEEE, Piscataway, 1998), pp. 264–269
15. V. Betz, J. Rose, How much logic should go in an FPGA logic block. IEEE Des. Test Comput. **15**(1), 10–15 (1998)
16. S. Ma, Y. Xu, W. Yue, Numerical solutions of a variable-order fractional financial system. J. Appl. Math. **2012**, Article ID 417942 (2012)

Chapter 6
Synchronization and Applications of Fractional-Order Chaotic Systems

6.1 Synchronizing Chaotic Oscillators in a Master-Slave Topology

The synchronization of two chaotic oscillators in a master-slave topology occurs when the trajectories of both state variables meet in the same time with a minimum synchronization error, so that they adjust their behavior temporarily. Synchronizing chaotic oscillators has been a challenge to guarantee successful applications in secure communications, which have been developed since the introduction of the first synchronization approach between two chaotic oscillators by Pecora and Carroll [1, 2]. Nowadays, several challenges remain open to accomplish and guarantee privacy and high security of the transmitted information, so that researchers are searching for the best integer/fractional-order chaotic/hyperchaotic oscillator and synchronization approaches.

During the last years, many synchronization techniques for fractional-order chaotic oscillators have been introduced along with some applications to random number generators, robotics, and security, for instance. The main objective of synchronizing two chaotic oscillators is oriented to develop secure communication systems to preserve privacy, provide security, and be invulnerable to attacks. These issues can be accomplished using chaotic oscillators because they have the property of high sensitivity to the initial conditions, which can be quantified by evaluating and maximizing the positive Lyapunov exponent, as described in Chap. 3. The evaluation of the fractal dimension also provides characteristics to rank the randomness and unpredictability of chaotic oscillators. In this manner, this Chapter describes the synchronization of fractional-order chaotic oscillators in a master-slave topology. It is also shown that for the fractional-order chaotic attractors having high positive Lyapunov exponents guarantees high security, and if the synchronization error is very low, then the original information can be recovered without loss of data. This issue is confirmed herein through evaluating the correlation among the original data, the chaotic channel that masks the information

© Springer Nature Switzerland AG 2020
E. Tlelo-Cuautle et al., *Analog/Digital Implementation of Fractional Order Chaotic Circuits and Applications*, https://doi.org/10.1007/978-3-030-31250-3_6

being transmitted with chaos, and the recovered data. To implement the fractional-order chaotic secure communication system, the Grünwald–Letnikov definition is applied, as described in Chaps. 2 and 5. The FPGA-based implementation of the fractional-order chaotic oscillators applying Grünwald–Letnikov method takes advantage of the short memory principle to reduce hardware resources. Based on the synchronization techniques of chaotic oscillators and their FPGA-based implementation for secure image transmission introduced in [3], Hamiltonian forms and Open-Plus-Closed-Loop (OPCL) synchronization techniques are applied herein.

6.1.1 Synchronizing Chaotic Oscillators in a "Master-Slave" Topology with Hamiltonian Forms and Observer Approach

Every integer-order chaotic oscillator can be described by $\dot{x} = f(x)$, which is according to the seminal work given in [4]; the Hamiltonian approach can be described by (6.1), where ∂H is the gradient vector of the energy function H, positive definite in R^n. H is a quadratic function defined by $H(x) = \frac{1}{2}X^T M x$, with M as a symmetrical matrix and positive definite. $J(x)$ and $S(x)$ are matrices representing the conservative and nonconservative parts of the system, respectively, and must satisfy: $J(x) + J^T(x) = 0$ and $S(x) = S^T(x)$. There exists the possibility to add a destabilizing vector as $F(x)$, to get the form of a Hamiltonian system, as shown in (6.2). This can consider suppositions to get the form given in (6.1), without $F(x)$.

$$\dot{x} = J(x)\frac{\partial H}{\partial x} + S(x)\frac{\partial H}{\partial x}, \quad x \in R^n \tag{6.1}$$

$$\dot{x} = J(x)\frac{\partial H}{\partial x} + S(x)\frac{\partial H}{\partial x} + F(x), \quad x \in R^n \tag{6.2}$$

If one considers the system with destabilizing vector and one linear output, one gets (6.3), where y is a vector denoting the output of the system. In addition, if ξ is the estimated state vector of x and η the estimated output in terms of ξ, then an observer to (6.2) can be given by (6.4), where K is a vector of constant gains.

$$\dot{x} = J(y)\frac{\partial H}{\partial x} + S(y)\frac{\partial H}{\partial x} + F(y), \quad x \in R^n$$

$$y = C\frac{\partial H}{\partial x}, \quad y \in R^m \tag{6.3}$$

$$\dot{\xi} = J(y)\frac{\partial H}{\partial \xi} + S(y)\frac{\partial H}{\partial \xi} + F(y) + K(y - \eta)$$

$$\eta = C\frac{\partial H}{\partial \xi}$$

(6.4)

The synchronization by Hamiltonian forms is achieved after accomplishing two theorems.

Theorem 6.1 *The state x of the nonlinear system (6.3) can be global, exponential, and asymptotically estimated by the state of an observer of the form (6.4), if the pair of matrices (C,S) are observables.*

Theorem 6.2 *The state x of the nonlinear system (6.3) can be global, exponential, and asymptotically estimated by the state of an observer of the form (6.4), if and only if there exists a constant matrix K such that the symmetric matrix in (6.5) be negative definite [4].*

$$[W - KC] + [W - KC]^T = [S - KC] + [S - KC]^T$$

$$= 2\left[S - \frac{1}{2}\left(KC + C^T K^T\right)\right]$$

(6.5)

6.1.2 Synchronizing Chaotic Oscillators in a "Master-Slave" Topology with OPCL

The Open-Plus-Closed-Loop (OPCL) technique is based on the control systems combination. It is a heterogeneous synchronization because it allows to obtain the master and slave parameters. From a dynamical system described by $\dot{x} = f(x)$, the master chaotic oscillator is given by (6.6), where $x_m(t)$, $y_m(t)$, and $z_m(t)$ denote the state variables, and then $x_e(t)$, $y_e(t)$, and $z_e(t)$ denote the slave chaotic oscillator in (6.7). $D(v(t), u(t))$ is given in (6.8), with D_1 and D_2 as open loop and closed loop parts, respectively, and given by (6.9) and (6.10).

$$\frac{d}{dt}u(t) = F(u(t)) = F(x_m(t), y_m(t), z_m(t)); \quad u \in \mathbb{R}^3$$

(6.6)

$$\frac{d}{dt}v(t) = F(v(t)) + D(v(t), u(t)); \qquad v \in \mathbb{R}^3$$

(6.7)

$$D(v(t), u(t)) = D_1(u(t)) + D_2(v(t), u(t));$$

(6.8)

$$D_1(u(t)) = \frac{du(t)}{dt} - F(u(t));$$

(6.9)

$$D_2(v(t), u(t)) = \left(H - \frac{\delta}{\delta t} F(u(t)) \right) e(t) \qquad (6.10)$$

H is an arbitrary constant Hurwitz matrix, so that the simplicity of the slave system depends on how this matrix is chosen. Besides, $e(t) = v(t) - u(t)$ is defined as synchronization error. For the OPCL synchronization to be achieved, the error must tend to zero and it can be verified by Taylor's series [5]. If the real parts of the eigenvalues from H are negative, the synchronization will be successful. This is a necessary condition but not enough since there may be an H with eigenvalues equal to zero and the synchronization can occur [6].

6.2   Synchronizing Fractional-Order Chaotic Oscillators

The synchronization techniques applied to integer-order chaotic oscillators produce a low error and both chaotic oscillators, the master and the slave, synchronize in very few iterations so that they are suitable for FPGA-based implementation. This section shows the synchronization of two fractional-order chaotic oscillators in a master-slave topology using as cases of study the Lorenz oscillator given in (6.11) and the Rössler oscillator given in (6.12). In both fractional-order chaotic systems, the Hamiltonian forms and OPCL synchronization techniques are applied.

$$\begin{aligned} D_t^{q_1} x &= \sigma(y - x), \\ D_t^{q_2} y &= x(\rho - z) - y, \\ D_t^{q_3} z &= xy - \beta z, \end{aligned} \qquad (6.11)$$

$$\begin{aligned} D_t^{q_1} x &= -(y + z), \\ D_t^{q_2} y &= x + ay, \\ D_t^{q_3} z &= b + z(x - c) \end{aligned} \qquad (6.12)$$

6.2.1   Hamiltonian Forms: Synchronization of Two Fractional-Order Lorenz Systems

Let us consider the fractional-order chaotic oscillator based on Lorenz equations, proposing the master system similar to the original one given in (6.11) with $\sigma = 10$, $\rho = 28$, and $\beta = 8/3$, and the energy function as in (6.13), then the Hamiltonian system given in (6.14) arises. It becomes the master and the slave system is proposed by adding the gain vector multiplied by the error. The gain vector is obtained verifying that it contains the pair of matrices (C, S). In this manner, the gain vector K can be obtained by applying the Sylvester criterion for negative definite matrices.

Herein the gains are equal to $k_1 = 8, k_2 = 8, k_3 = 3$ and the observer system is described by (6.15). Finally, the slave system is given in (6.16),

$$H(x) = \frac{1}{2}[x^2 + y^2 + z^2] \tag{6.13}$$

$$\begin{bmatrix} x_m \\ y_m \\ z_m \end{bmatrix} = \begin{bmatrix} 0 & -9 & 0 \\ -9 & 0 & 0 \\ 0 & 0 & 0 \end{bmatrix} \frac{\partial H}{\partial x} + \begin{bmatrix} -10 & 19 & 0 \\ 19 & -1 & 0 \\ 0 & 0 & -2.667 \end{bmatrix} \frac{\partial H}{\partial x} + \begin{bmatrix} 0 \\ -x_m * z_m \\ x_m * y_m \end{bmatrix} \tag{6.14}$$

$$\begin{bmatrix} x_e \\ y_e \\ z_e \end{bmatrix} = \begin{bmatrix} 0 & -9 & 0 \\ -9 & 0 & 0 \\ 0 & 0 & 0 \end{bmatrix} \frac{\partial H}{\partial x} + \begin{bmatrix} -10 & 19 & 0 \\ 19 & -1 & 0 \\ 0 & 0 & -2.667 \end{bmatrix} \frac{\partial H}{\partial x} + \begin{bmatrix} 0 \\ -x_e * z_e \\ x_e * y_e \end{bmatrix}$$
$$+ \begin{bmatrix} 8 \\ 8 \\ 3 \end{bmatrix} (y - \eta) \tag{6.15}$$

$$\begin{aligned} D_t^{q_1} x_e &= \sigma(y_e - x_e) + 8[x_m - x_e], \\ D_t^{q_2} y_e &= x_e(\rho - z_e) - y_e + 8[y_m - y_e], \\ D_t^{q_3} z_e &= x_e y_e - \beta z_e + 3[z_m - z_e], \end{aligned} \tag{6.16}$$

The synchronization among the state variables of the master and slave systems is shown in the phase diagrams in Fig. 6.1. Figure 6.2 shows the attractor of the fractional-order Lorenz oscillator, first when the synchronization does not occur and then when it successfully occurs applying the Hamiltonian forms and observer approach technique. The time series of the master and slave fractional-order chaotic systems are shown in Fig. 6.3. The synchronization error between the master and the slave fractional-order chaotic systems is shown in Fig. 6.4, where it can be seen that the synchronization is accomplished around iteration 350. Thinking on an FPGA-based implementation with a clock CLK above MHz, this delay is very small to be appreciated in a real application for security, as shown in the following section.

6.2.2 OPCL: Synchronization of Two Fractional-Order Lorenz Systems

The OPCL synchronization technique is also applied herein considering two fractional-order Lorenz chaotic systems. In this synchronization technique, the master fractional-order system is proposed to be similar to the original system. The open loop part in the fractional-order slave system is null ($D_1(u(t)) = 0$). For the

Fig. 6.1 Phase-space
diagrams for the master and
slave state variables: (**a**) x_m,
x_s; (**b**) y_m, y_s; and (**c**) z_m, z_s
for the fractional-order
Lorenz system applying
Hamiltonian forms and
observer approach

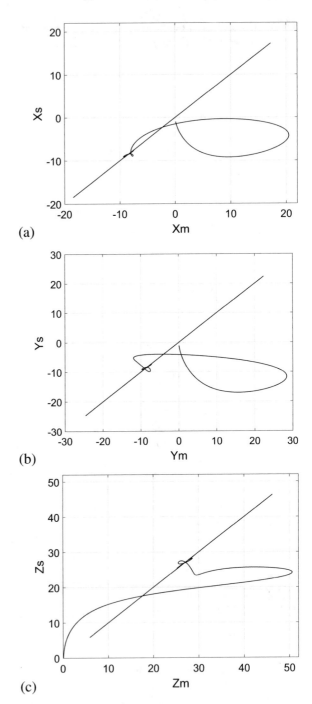

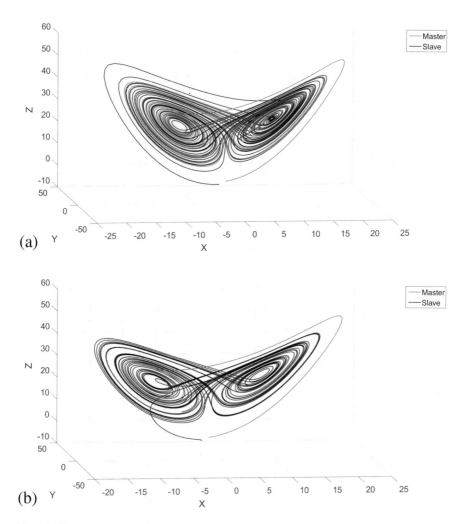

Fig. 6.2 Master and slave phase-space portraits in 3D of the fractional-order Lorenz system applying Hamiltonian forms: (**a**) before and (**b**) after synchronization occurs

closed part, the fractional-order master system partial derivative is given in (6.17), and H is proposed in (6.18), where P is a constant value and depending on how many values are proposed, it will be the complexity to obtain the closed loop part. The eigenvalues of H determine that p_1 and p_2 must be negative. For example, if $p_1 = -10$ and $p_2 = -10$, the eigenvalues shown in (6.19) have real part negative and thereby the condition described above is accomplished. Therefore, the closed loop contribution is given in (6.20). Finally, with the open-closed loop contribution, the fractional-order chaotic slave oscillator for Lorenz system is given in (6.21).

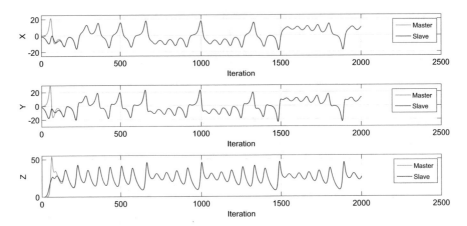

Fig. 6.3 Time series of the master and slave fractional-order Lorenz systems applying Hamiltonian forms

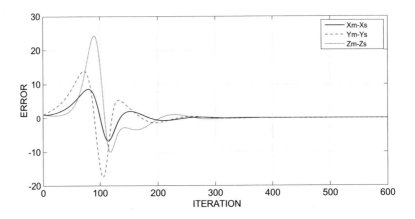

Fig. 6.4 Synchronization error of the fractional-order Lorenz systems, for the master and slave state variables applying Hamiltonian forms

$$\frac{\delta}{\delta t} F(u(t)) = \begin{pmatrix} -\sigma & \sigma & 0 \\ \rho - z_m & -1 & -x_m \\ y_m & x_m & -\beta \end{pmatrix} \tag{6.17}$$

$$H = \begin{pmatrix} -\sigma + p_1 & \sigma + p_2 & 0 \\ \rho & -1 & 0 \\ 0 & 0 & -\beta \end{pmatrix} \tag{6.18}$$

$$\lambda_1 = -2.6667$$

$$\lambda_2 = -20.0 \tag{6.19}$$

$$\lambda_3 = -1.0$$

$$
D_2 = \left(\begin{pmatrix} -\sigma + p_1 & \sigma + p_2 & 0 \\ \rho & -1 & 0 \\ 0 & 0 & -\beta \end{pmatrix} - \begin{pmatrix} -\sigma & \sigma & 0 \\ \rho - z_m & -1 & -x_m \\ y_m & x_m & -\beta \end{pmatrix} \right) (v_t - u_t)
$$

$$
= \begin{pmatrix} p_1(x_e - x_m) + p_2(y_e - y_m) \\ z_m(x_e - x_m) + x_m(z_e - z_m) \\ -y_m(x_e - x_m) - x_m(y_e - y_m) \end{pmatrix} \tag{6.20}
$$

$$
\begin{aligned}
D_t^{q_1} x_e &= \sigma(y_e - x_e) + p_1(x_e - x_m) + p_2(y_e - y_m), \\
D_t^{q_2} y_e &= x_e(\rho - z_e) - y_e + z_m(x_e - x_m) + x_m(z_e - z_m), \\
D_t^{q_3} z_e &= x_e y_e - \beta z_e - y_m(x_e - x_m) - x_m(y_e - y_m)
\end{aligned} \tag{6.21}
$$

With $p_1 = -10$ and $p_2 = -10$, the synchronization between the master and slave fractional-order chaotic oscillators occur, and it is shown in the phase-space diagrams in Fig. 6.5. The synchronization errors are shown in Fig. 6.6, where it is observed that the minimum error occurs around iteration 750.

6.2.3 Hamiltonian Forms: Synchronization of Two Fractional-Order Rössler Systems

As showed before for the synchronization of two fractional-order Lorenz systems applying Hamiltonian forms and observer approach, a similar approach is performed for the fractional-order chaotic oscillator based on Rössler equations. Again, the first step consists of proposing a fractional-order chaotic master system similar to the original one given in (6.12), which energy function can be proposed as given in (6.22). Then the Hamiltonian system given in (6.23) arises. It becomes the fractional-order master and the slave chaotic system, which is proposed by adding the gain vector multiplied by the error. In this case, the gain vector is obtained verifying that it contains the pair of matrices (C, S). In this manner, the gain vector K can be obtained by applying the Sylvester criterion for negative definite matrices. For the case of using the fractional-order Rössler system, herein the gains are equal to $k_1 = 2, k_2 = 1, k_3 = 4$ and the fractional-order chaotic observer system can be described by (6.24). Finally, the fractional-order chaotic slave system is given in (6.25),

$$H(x) = \frac{1}{2}\left[x^2 + y^2 + z^2 \right] \tag{6.22}$$

Fig. 6.5 Phase-space diagrams for the master and slave state variables: (**a**) x, (**b**) y, and (**c**) z for the fractional-order Lorenz system applying OPCL technique

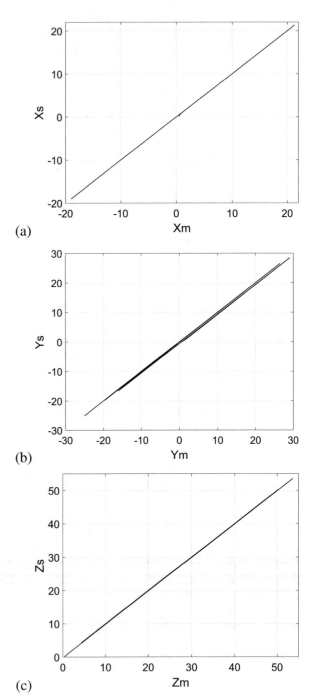

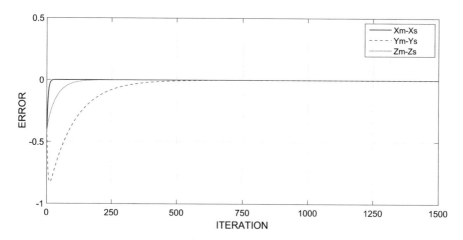

Fig. 6.6 Synchronization error of the fractional-order Lorenz systems, for the master and slave state variables applying OPCL technique

$$
\begin{bmatrix} x_m \\ y_m \\ z_m \end{bmatrix} = \begin{bmatrix} 0 & -1 & -0.5 \\ 1 & 0 & 0 \\ 0.5 & 0 & 0 \end{bmatrix} \frac{\partial H}{\partial x} + \begin{bmatrix} 0 & 0 & -0.5 \\ 0 & 0.5 & 0 \\ -0.5 & 0 & -10 \end{bmatrix} \frac{\partial H}{\partial x} + \begin{bmatrix} 0 \\ 0 \\ x_m z_m + 0.2 \end{bmatrix}
$$
(6.23)

$$
\begin{bmatrix} x_e \\ y_e \\ z_e \end{bmatrix} = \begin{bmatrix} 0 & -1 & -0.5 \\ 1 & 0 & 0 \\ 0.5 & 0 & 0 \end{bmatrix} \frac{\partial H}{\partial x} + \begin{bmatrix} 0 & 0 & -0.5 \\ 0 & 0.5 & 0 \\ -0.5 & 0 & -10 \end{bmatrix} \frac{\partial H}{\partial x} + \begin{bmatrix} 0 \\ 0 \\ x_e z_e + 0.2 \end{bmatrix}
$$
$$
+ \begin{bmatrix} 2 \\ 1 \\ 4 \end{bmatrix} (y - \eta)
$$
(6.24)

$$
\begin{aligned}
D_t^{q_1} x_e &= -(y_e + z_e) + 2[x_m - x_e], \\
D_t^{q_2} y_e &= x_e + a y_e + [y_m - y_e], \\
D_t^{q_3} z_e &= b + z(x_e - c) + 4[z_m - z_e],
\end{aligned}
$$
(6.25)

The synchronization among the state variables of the fractional-order chaotic master and slave systems is shown in the phase-space diagrams shown in Fig. 6.7. Figure 6.8 shows the fractional-order chaotic attractor of the Rössler oscillator, first when the synchronization does not occur and then when it successfully occurs when applying the Hamiltonian forms and observer approach technique. The fractional-order chaotic time series of the master and slave systems is shown in Fig. 6.9. The

Fig. 6.7 Phase-space diagrams for the fractional-order chaotic master and slave state variables: (**a**) x, (**b**) y, and (**c**) z for Rössler system applying Hamiltonian forms and observer approach

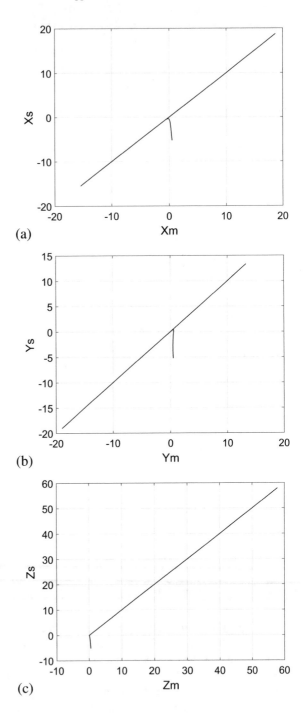

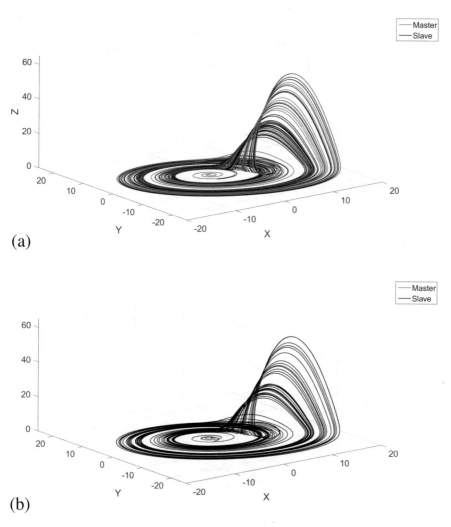

Fig. 6.8 Master and slave phase-space portraits in 3D of the fractional-order Rössler chaotic system applying Hamiltonian forms: (**a**) before and (**b**) after synchronization occurs

synchronization error between the master and the slave fractional-order chaotic systems is shown in Fig. 6.10, where it can be seen that the synchronization is accomplished around iteration 200.

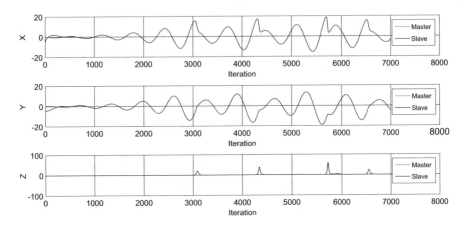

Fig. 6.9 Time series of the master and slave fractional-order Rössler systems applying Hamiltonian forms

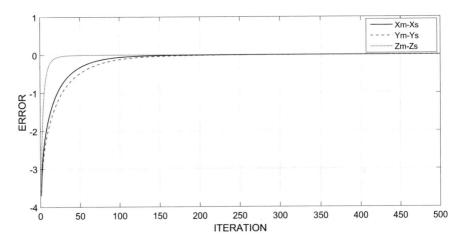

Fig. 6.10 Synchronization error of the fractional-order Rössler chaotic systems, for the master and slave state variables applying Hamiltonian forms

6.2.4   OPCL: Synchronization of Two Fractional-Order Rössler Systems

Applying the OPCL synchronization technique, let us consider the fractional-order Rössler chaotic system with $a = 0.5$, $b = 0.2$, and $c = 10$; and the fractional-order chaotic master system is proposed being similar to the original one. The open loop part in the fractional-order chaotic slave system is null $(D_1(u(t)) = 0)$. For the closed part, the fractional-order chaotic master system partial derivative is given in (6.26), and H is proposed in (6.27), where P is a constant value and depending on how many values are proposed, it will be the complexity to obtain the closed loop part. The eigenvalues of H determine that p_1 and p_2 must be negative. For example, if $p_1 = -1$, the eigenvalues shown in (6.28) have real part negative and thereby the condition described above is accomplished. Therefore, the closed loop contribution is given in (6.29). Finally, with the open-closed loop contribution, the fractional-order chaotic slave oscillator for Rössler system is given in (6.30).

$$\frac{\delta}{\delta t} F(u(t)) = \begin{pmatrix} 0 & -1 & -1 \\ 1 & a & 0 \\ z_m & 0 & x_m - c \end{pmatrix} \tag{6.26}$$

$$H = \begin{pmatrix} p_1 & -1 & -1 \\ 1 & a + p_2 & 0 \\ 0 & 0 & -c \end{pmatrix} \tag{6.27}$$

$$\begin{aligned} \lambda_1 &= -10.0 \\ \lambda_2 &= -1.7500 - 0.9682i \\ \lambda_3 &= -1.7500 + 0.9682i \end{aligned} \tag{6.28}$$

$$\begin{aligned} D_2 &= \left(\begin{pmatrix} p_1 & -1 & -1 \\ 1 & a + p_2 & 0 \\ 0 & 0 & -c \end{pmatrix} - \begin{pmatrix} 0 & -1 & -1 \\ 1 & a & 0 \\ z_m & 0 & x_m - c \end{pmatrix} \right) (v_t - u_t) \\ &= \begin{pmatrix} p_1(x_e - x_m) \\ p_2(y_e - y_m)) \\ -z_m(x_e - x_m) - x_m(z_e - z_m) \end{pmatrix} \end{aligned} \tag{6.29}$$

$$\begin{aligned} D_t^{q_1} x_e &= -(y_e + z_e) + p_1(x_e - x_m), \\ D_t^{q_2} y_e &= x_e + a y_e + p_2(y_e - y_m), \\ D_t^{q_3} z_e &= b + z_e(x_e - c) - z_m(x_e - x_m) - x_m(z_e - z_m), \end{aligned} \tag{6.30}$$

With $p_1 = -2$ and $p_2 = -2$ the synchronization between the fractional-order master and slave chaotic oscillators occurs, and it is shown in the phase-space diagrams in Fig. 6.11. The synchronization errors between the two fractional-order chaotic attractors are shown in Fig. 6.12, where it is observed that the minimum error occurs around iteration 750.

Fig. 6.11 Phase-space diagrams for the master and slave state variables: (**a**) x, (**b**) y, and (**c**) z for the fractional-order Rössler chaotic system applying OPCL technique

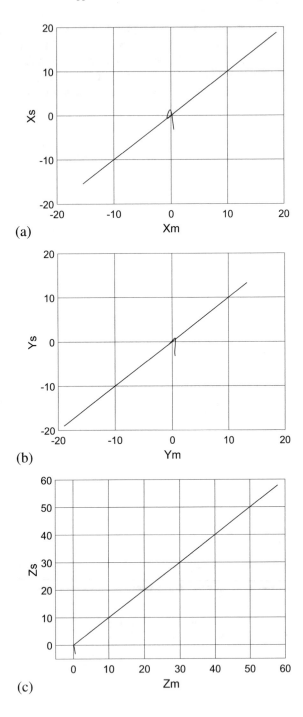

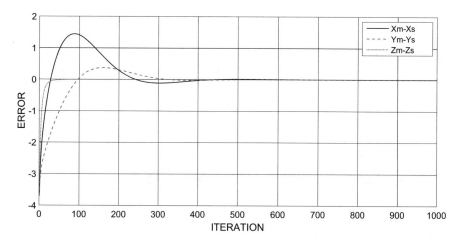

Fig. 6.12 Synchronization error of the fractional-order Rössler chaotic systems, for the master and slave state variables applying OPCL technique

6.3 Fractional-Order Chaotic Secure Communication System Applied to Image Encryption

Using the synchronization approaches described above, i.e., Hamiltonian forms and observer approach, and OPCL techniques, one can implement a fractional-order chaotic secure communication system. The system is sketched in Fig. 6.13. It can be appreciated that this secure communication system has an image as input, which is masking with the chaotic time series generated by the fractional-order chaotic oscillator in the master block. The channel is then transmitting an encrypted image, which can be recovered in the slave block that contains another fractional-order chaotic oscillator that is synchronized with the one in the master block to generate the same chaotic sequence and then to subtract the chaotic data. At the end, for a synchronization that generates the minimum error between two fractional-order chaotic oscillators, the original image is recovered at the slave block.

Figure 6.14 shows the experimental results of the masking of the cameraman black&white 256×256 pixels image. In this case the data is processed using MATLAB and applying the Hamiltonian forms and observer approach of the master-slave synchronization of two fractional-order Lorenz chaotic oscillators. The masking is performed for each state variable used to mask the original image. In this manner, each column in Fig. 6.14 represents the original image, the masking process, and the recovered image. Each row is associated with each state variable of the fractional-order chaotic oscillator, so that one can appreciate that not all of them are suitable for chaotic masking. The same process is performed using the fractional-order Rössler chaotic oscillator and also both Hamiltonian forms and OPCL synchronization techniques. All the simulation results are summarized in Table 6.1, which shows the correlations between the original image and the chaotic

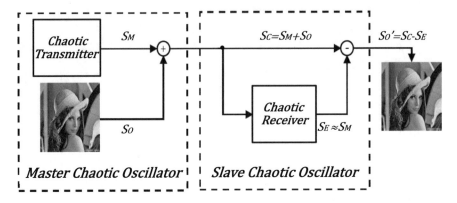

Fig. 6.13 Secure communication system based on the synchronization of two fractional-order chaotic oscillator in a master-slave topology

Fig. 6.14 Original (first column), encrypted (second column), and recovered (third column) images when applying Hamiltonian forms for the synchronization of the fractional-order Lorenz chaotic oscillator, to perform masking to the cameraman image using the chaotic time series of the state variable: (**a**) x in the first row, (**b**) y in the second row, and (**c**) z in the third row

Table 6.1 Correlations among the original, encrypted, and recovered data applying Hamiltonian forms and OPCL techniques to the fractional-order Lorenz and Rössler chaotic oscillators, and masking Cameraman's image with each state variable x, y, z

Synchronization technique	Chaotic oscillator	Transmission variable	Transmission correlation $\frac{\text{Original img.+Chaotic signal}}{\text{Original img.}}$	Recovery correlation $\frac{\text{Original img.}}{\text{Recovery img.}}$
Hamiltonian forms	Lorenz	x	0.0060	1
		y	0.0033	1
		z	7.9590e−05	1
	Rössler	x	0.0085	1
		y	0.0012	1
		z	0.0059	1
OPCL	Lorenz	x	0.0021	1
		y	0.0040	1
		z	0.0010	1
	Rössler	x	0.0036	1
		y	0.0019	1
		z	0.0202	1

channel, and between the original and the recovered images. According to the correlation analyses between the original image and the chaotic channel, the best synchronization technique that generates the lowest correlation is the one based on Hamiltonian forms and when using the state variable z of the fractional-order Lorenz chaotic oscillator.

Figure 6.15 shows the experimental results of the Lena RGB 512×512 pixels image when it is transmitted in a synchronized system applying Hamiltonian forms to the fractional-order Lorenz chaotic oscillator. The masking is performed using the three state variables, so that one can observe the original image in the first column, the masked image in the second column, and the recovered image in the last column. In the rows, one can see the cases when using the state variable x to mask Lena's image on the first row, using y in the second row, and using z in the third row. Again, applying both Hamiltonian forms and OPCL synchronization techniques, to both fractional-order Lorenz and Rössler chaotic oscillator, and using the three state variables to perform the masking operation, the correlation results are given in Table 6.2, showing the correlations between the original image and the chaotic channel, and between the original and the recovered images. According to the correlation analyses between the original image and the chaotic channel, the best synchronization technique in this case is the one based on OPCL when using the state variable y of the fractional-order Rössler chaotic oscillator.

Figure 6.16 shows the experimental results of the encryption of the Baboon RGB 512×512 pixels image, when applying Hamiltonian forms to the Fractional-order chaotic oscillator based on Lorenz equations. Similar to the previous examples, the secure transmission is performed by using the three state variables to observe the original image in the first column, the masked image in the second column, and the recovered image in the last column. The simulation results are summarized

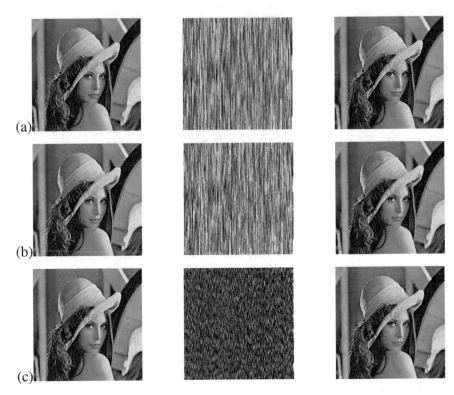

Fig. 6.15 Original (first column), encrypted (second column), and recovered (third column) images applying Hamiltonian forms to the fractional-order chaotic oscillator based on Lorenz equations, and using the state variable: (**a**) x in the first row, (**b**) y in the second row, and (**c**) z in the third row to mask Lena's image

in Table 6.3, showing the correlations between the original image and the chaotic channel, and between the original and the recovered images. According to the correlation between the original image and the chaotic channel, the best synchronization technique is the one based on Hamiltonian forms, and using the state variable x of the fractional-order Lorenz chaotic oscillator to perform the masking process.

6.4 FPGA-Based Implementation of a Secure Communication System Based on Fractional-Order Chaotic Oscillators

The VHDL hardware description and implementation of the secure communication system based on fractional-order chaotic oscillators are developed herein using two FPGA boards Cyclone IV GX EP4CGX150DF31C7 and a personal computer. The

Table 6.2 Correlations among the original, encrypted, and recovered data applying Hamiltonian forms and OPCL techniques to the fractional-order Lorenz and Rössler chaotic oscillators, and masking Lena's image with each state variable x, y, z

Synchronization technique	Chaotic oscillator	Transmission variable	Transmission correlation $\frac{\text{Original img.+Chaotic signal}}{\text{Original img.}}$	Recovery correlation $\frac{\text{Original img.}}{\text{Recovery img.}}$
Hamiltonian forms	Lorenz	x	0.0088	1
		y	0.0081	1
		z	0.0075	1
	Rössler	x	0.0024	1
		y	0.0043	1
		z	0.0020	1
OPCL	Lorenz	x	0.0038	1
		y	0.0039	1
		z	0.0021	1
	Rössler	x	0.0027	1
		y	5.1140e−04	1
		z	0.0090	1

image being transmitted in a real secure communication system can be of grayscale or RGB type. The image can be send to the FPGA-based chaotic secure communication system by a personal computer through the serial RS-232 communication protocol. Figure 6.17 shows the block diagram of the interconnections among the personal computer and two FPGAs to implement the chaotic secure communication system, and it can be associated with the secure communication system based on the synchronization of two fractional-order chaotic oscillator in a master-slave topology shown in Fig. 6.13. In Fig. 6.17 the FPGA1 contains the transmission part and the finite state machine (FSM1 block) that controls this transmitter block. The signal EOS activates the secure communication system when the master-slave synchronization (fractional-order chaotic oscillator's blocks) is successful and the error is the lowest one. Then signal REN enables the serial communication through RS-232 protocol in the receiver block. At this time, any type of data can be received serially by the secure communication system and when the reception finishes, signal EOR is activated. Then, FSM enables the adder block, which is in charge of masking the image with the chaotic data, with the AEN signal and sends the encrypted data (original image + chaos) to the receiver block that is implemented on the FPGA2, and which contains the receiver (slave fractional-order chaotic oscillator) block that is controlled by the finite state machine (FSM2 block), and is enabled with the FEN signal. When the encrypted data is in the receiver block in the PFGA2, the subtractor block is activated with SEN signal, which is responsible for recovering the original data by subtracting the chaotic data. Finally, the multiplexer block is enabled with signal MC to transmit the encrypted and recovered data through the RS-Transmitter block controlled by the TEN signal to the personal computer. When the transmission

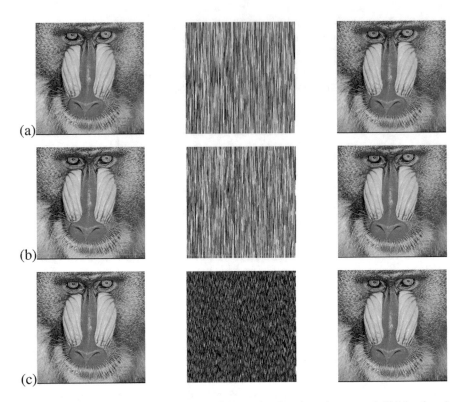

Fig. 6.16 Original (first column), encrypted (second column), and recovered (third column) images applying Hamiltonian forms to the fractional-order chaotic oscillator based on Lorenz equations, and using the state variable: (**a**) x in the first row, (**b**) y in the second row, and (**c**) z in the third row to mask Baboon's image

finishes, signal EOT is activated and the secure communication system is ready to receive new data.

Figures 6.18 and 6.19 show the experimental phase-space diagrams of the synchronization of fractional-order chaotic oscillators implemented on FPGA, using Lorenz and Rössler equations, respectively. It can be noted that the synchronization error is very low in all cases, so that they are suitable for implementing a secure communication system.

Figure 6.20 shows the experimental results of the secure transmission using the system shown in Fig. 6.17. The three images described above are taken to perform the encryption operation. One can observe the original image in the first column, the encrypted image in the second column, and the recovered image in the last column. The experimental results are summarized in Table 6.4, showing the correlations between the original image and the chaotic channel, and between the original and the recovered images. As one sees, the recovered image is the same as the original

Table 6.3 Correlations among the original, encrypted, and recovered data applying Hamiltonian forms and OPCL techniques to the fractional-order Lorenz and Rössler chaotic oscillators, and masking Baboon's image with each state variable x, y, z

Synchronization technique	Chaotic oscillator	Transmission variable	Transmission correlation $\frac{\text{Original img.+Chaotic signal}}{\text{Original img.}}$	Recovery correlation $\frac{\text{Original img.}}{\text{Recovery img.}}$
Hamiltonian forms	Lorenz	x	1.7374e−04	1
		y	6.0532e−04	1
		z	0.0062	1
	Rössler	x	0.0020	1
		y	3.7533e−04	1
		z	2.0639e−04	1
OPCL	Lorenz	x	0.0016	1
		y	0.0011	1
		z	0.0013	1
	Rössler	x	0.0052	1
		y	0.0012	1
		z	0.0051	1

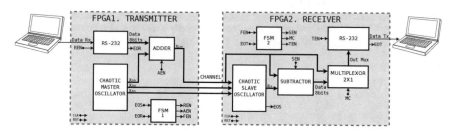

Fig. 6.17 Sketching the implementation of a chaotic secure communication system based on a master-slave synchronized topology and using fractional-order chaotic oscillators

one and the correlations between the original and the masked images are very low, thus showing the usefulness of the scheme shown in Fig. 6.17 for image encryption.

6.5 Fractional-Order Chaos-Based Random Number Generators

The generation of random numbers based on chaotic oscillators has been a challenge to guarantee the best randomness. Fractional-order chaotic oscillators can also be used to implement a random number generator, which can be tested to probe that it is suitable for cryptographic applications. Let us consider the fractional-order Lorenz chaotic oscillator, and its state variable z, which is implemented on an FPGA applying the Grünwald–Letnikov method. The chaotic time series is shown in

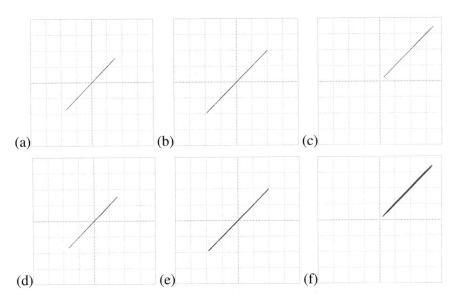

Fig. 6.18 Experimental phase-space diagrams for the master (horizontal axis) vs slave (vertical axis) state variables: (**a**) x, (**b**) y, and (**c**) z for synchronizing the fractional-order Lorenz systems applying Hamiltonian forms and observer approach, and (**d**) x, (**e**) y, and (**f**) z applying OPCL synchronization technique implemented on an FPGA, and observed with 1 V/Div in a Lecroy oscilloscope

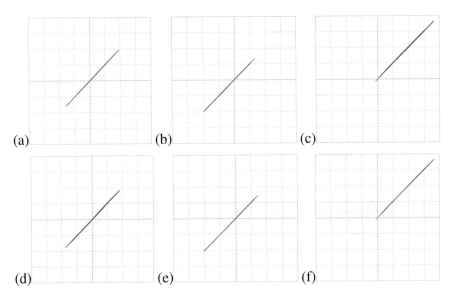

Fig. 6.19 Experimental phase-space diagrams for the master (horizontal axis) vs slave (vertical axis) state variables: (**a**) x, (**b**) y, and (**c**) z for synchronizing the fractional-order Rössler systems applying Hamiltonian forms and observer approach, and (**d**) x, (**e**) y, and (**f**) z applying OPCL synchronization technique implemented on an FPGA, and observed with 1 V/Div in a Lecroy oscilloscope

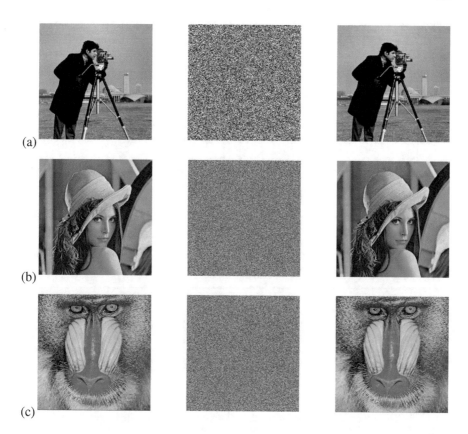

Fig. 6.20 Original (first column), encrypted (second column), and recovered (third column) images processed through the secure communication system shown in Fig. 6.17 and applying Hamiltonian forms to the fractional-order Lorenz chaotic oscillator using the state variable z: (**a**) Cameraman, (**b**) Lena, and (**c**) Mandril

Table 6.4 Correlations among the original, encrypted, and recovered data using three images and the secure communication system shown in Fig. 6.17 and applying Hamiltonian forms to the fractional-order Lorenz chaotic oscillator using the state variable z

Original image	Transmission variable	Transmission correlation $\frac{\text{Original img.+Chaotic signal}}{\text{Original img.}}$	Recovery correlation $\frac{\text{Original img.}}{\text{Recovery img.}}$
Cameraman	z	0.0081	1
Lena	z	8.7764e−04	1
Baboon	z	0.0010	1

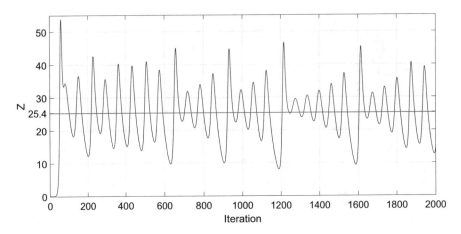

Fig. 6.21 State variable z and its chaotic time series from the fractional-order Lorenz chaotic oscillator

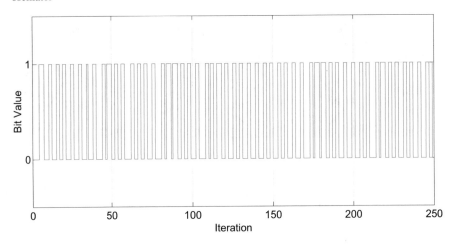

Fig. 6.22 Bitstream obtained from the chaotic time series of Fig. 6.21, with a threshold established at 25.4

Fig. 6.21. Using this data, the task is choosing an appropriate threshold to generate bitstream which number of logic 1s and logic 0s be the most close to 50% in each case. A comparator can be implemented to perform this task, and then a value above the selected threshold will generate a bit classified as logic 1; otherwise, the generated bit is classified as logic 0. In this manner, Fig. 6.22 shows the generated bitstream considering the first 250 samples from the chaotic time series given in Fig. 6.21, in which the threshold was set to 25.4 so that the percentage of logic 1s is equal to 50.1% and the percentage of logic 0 is equal to 49.9%. It is worth mentioning that when the threshold is set to 26.5, the percentages are 43.2% for logic 1s and 56.8% for logic 0s. Also, when the threshold is set to 24.5, i.e., above the threshold of 25.4, the percentages are 54.25% for logic 1s and 46.75% for logic 0s, thus confirming that the selection of the best threshold matters to improve the randomness.

References

1. L.M. Pecora, T.L. Carroll, Synchronization in chaotic systems. Phys. Rev. Lett. **64**(8), 821 (1990)
2. L.M. Pecora, T.L. Carroll, Synchronizing chaotic circuits. IEEE Trans. Circuits Syst. **38**, 453–456 (1991)
3. O. Guillén-Fernández, A. Meléndez-Cano, E. Tlelo-Cuautle, J.C. Núñez-Pérez, J. de Jesus Rangel-Magdaleno, On the synchronization techniques of chaotic oscillators and their FPGA-based implementation for secure image transmission. PloS One **14**(2), e0209618 (2019)
4. H. Sira-Ramirez, C. Cruz-Hernández, Synchronization of chaotic systems: a generalized Hamiltonian systems approach. Int. J. bifurcation Chaos **11**(5), 1381–1395 (2001)
5. E.A. Jackson, I. Grosu, An open-plus-closed-loop (OPCL) control of complex dynamic systems. Phys. D Nonlinear Phenomena **85**(1–2), 1–9 (1995)
6. A.I. Lerescu, N. Constandache, S. Oancea, I. Grosu, Collection of master–slave synchronized chaotic systems. Chaos Solitons Fractals **22**(3), 599–604 (2004)

Chapter 7
Conclusions and Issues for Future Research

Chaos is a hot topic in electronics engineering because it finds applications in robotics, random number generators, secure communications systems, security for Internet of Things (IoT), and so on. The simulation and analog/digital implementation of integer-order chaotic oscillators have been matured in the last years and some realizations have been performed using integrated circuit technology, which are very low power and voltage consumption. Those integrated circuits can be embedded into sensors, actuators, and processing circuits to encrypt any kind of data.

Integer-order chaotic and hyperchaotic oscillators can also be programmed in micro-controllers like Arduino or Raspberry Pi to encrypt data. One can take advantage of its capabilities to be connected to internet so that they can find applications in IoT. For example, they can be programmed to connect to MQTT protocol, in a machine-to-machine (M2M) communication topology. That way, a huge number of applications can be developed, like the ones listed in https://mqtt. org, such as: in sensors communicating to a broker via satellite link, over occasional dial-up connections with healthcare providers, and in a range of home automation and small device scenarios. MQTT is also ideal for mobile applications because of its small size, low power usage, minimized data packets, and efficient distribution of information to one or many receivers. In this regard, the chaotic oscillators presented herein can be either programmed in mobile or embedded systems, or interconnected to electronic systems for security and cryptographic applications if they are implemented with integrated circuit technology or electronic devices. This is a challenge to demonstrate that chaotic and hyperchaotic oscillators can compete with cryptographic hash functions already embedded on Arduino's micro-controller, or to generate true and pseudo-random number generators.

The chaotic oscillators described herein can be analyzed to evaluate their equilibrium points and eigenvalues, as shown in Chap. 1. However, they can be more easy to attack in secure applications, so that the new direction is the study and implementation of chaotic oscillators having hidden attractors, as recently discussed in [1]. In this manner, self-excited attractors have a basin of attraction

© Springer Nature Switzerland AG 2020
E. Tlelo-Cuautle et al., *Analog/Digital Implementation of Fractional Order Chaotic Circuits and Applications*, https://doi.org/10.1007/978-3-030-31250-3_7

that is excited from an unstable equilibrium point, as the classical nonlinear systems such as Lorenz's, Rössler's, Chen's, Lü's, or Sprott's systems. The modern focus on chaos theory is the study of hidden attractors, for which one must develop specific computational procedures to identify them because the evaluation of their equilibrium points is difficult. In the same direction are the fractional-order chaotic oscillators, for which still more analysis is required to introduce better approximations of their solutions and then develop improved analog/digital implementations.

In the previous chapters, some recent contributions on the optimization, control, circuit realizations, and applications of fractional-order systems were discussed. Issues on selecting the more appropriate numerical method are still under investigation to diminish numerical errors and speed-up time simulation, which is in direct relation to the hardware implementation. Analog implementations can be enhanced using FPAAs to reduce mismatches when using commercially available amplifiers, and also one can infer that the design of integrated circuits is a challenge to develop lightweight cryptographic applications, suitable for hardware security for IoT. On the other hand, FPGA-based implementations are just to verify the functionality of novel chaotic oscillators, which architecture can be further implemented on a system on chip like Raspberry Pi, or into a digital integrated circuit. One can also infer that the electronic implementation of fractional-order chaotic oscillators is more challenging due to the lack of robust approximations of their solution in either frequency or time domains.

Chapter 2 described the implementation of chaotic oscillators using FPAAs, which are analog signal processors, equivalent to the digital ones like FPGAs. Both can be reconfigurable electrically, with the advantage that an FPAA can be used to implement a wide variety of analog functions, such as integration, derivation, pondered sum/subtraction, filtering, rectification, comparator, multiplication, division, analog-to-digital conversion, voltage references, signal conditioning, amplification, synthesis of nonlinear functions, and generation of arbitrary signals, among others. On the other hand, FPGAs are programmable semiconductor devices that are based on a matrix of configurable logic blocks (CLBs) connected through programmable interconnects. Modern FPGAs include special hardware blocks like random access memory (RAM), which were used in the implementations of fractional-order chaotic oscillators in Chap. 5. However, one must be aware that the length of the digital word to represent real numbers matters. Chapter 2 detailed the effect of truncating bits, which must be performed carefully to capture the dominant behavior of the chaotic attractors.

Having a huge number of integer-order, hyperchaotic, and fractional-order chaotic oscillators, one is interested in choosing the ones that generate more unpredictability of the time series data. For this problem, one can evaluate the Lyapunov exponents, fractal dimension, and entropy, which can be combined to propose a figure of merit and then find the best chaos generators to enhance the applications described above. Chapter 3 showed that those characteristics can be optimized, and that this process is more suitable to apply metaheuristics due to the huge search spaces that arises for the design variables.

The implementation of fractional-order chaotic oscillators requires the use of fractional-order integrators to solve the fractional derivatives. These issues were described in Chap. 4 applying frequency domain methods, which approximate the fractional-order derivatives by Laplace transfer functions. Those functions can be implemented using fractors or fractances in different topologies that combine amplifiers and RC circuit elements. Also, the transfer functions can be implemented on FPAAs. But one must down-scale the mathematical models to generate values in the range that the electronic devices can drive. This problem does not exist using FPGAs, in which the fractional-order chaotic oscillators are simulated in time domain using different approximations like Grünwald-Letnikov and Adams-Bashforth-Moulton methods. Those methods were used to discretize the mathematical models of several fractional-order chaotic attractors, and to describe them in VHDL description language in Chap. 5. In that chapter it was also detailed the fixed-point representation of the numerical values and its two complement notations. An important thing is that using the floating point representation requires more hardware resources on an FPGA, so that fixed-point representation of real numbers is much better. In addition, it provides higher speed and lower cost of hardware implementation. The challenges are related to reach high throughputs and low latency of the FPGA-based implementations of the fractional-order chaotic attractors.

One of the very first applications of chaotic oscillators was the implementation of chaotic secure communication systems, which were based on applying Pecora and Carroll synchronization technique [2, 3]. Nowadays, several challenges remain open to accomplish and guarantee privacy and high security of the transmitted information, so that researchers are searching for the best integer/fractional-order chaotic/hyperchaotic oscillator and more robust synchronization approaches. For instance, Chap. 6 detailed the synchronization of two fractional-order chaotic oscillators in a master-slave topology. The FPGA-based implementation was extended to design a secure communication system which was used to transmit several images, and the correlation among the original images with the chaotic channel and the recovered images were presented. The results showed that for fractional-order chaotic oscillators, one must test all state variables to find the best one that better mask the transmitted data. The details on the synthesis of the architecture of a fractional-order-based chaotic secure communication system was detailed as well as the use of different memories, and the design of the control unit as a finite state machine. From the FPGA-based implementation of a fractional-order chaotic oscillator, Chap. 6 showed the implementation of a random number generator, for which the monobit test was highlighted counting the number of logic "1" and logic "0" when varying a threshold to generate binary strings. This is also a challenge to design true and pseudo-random number generators that can compete with other approaches like using physically unclonable functions, ciphers, and so on.

References

1. V.-T. Pham, S. Vaidyanathan, C. Volos, T. Kapitaniak, *Nonlinear Dynamical Systems with Self-excited and Hidden Attractors*, vol. 133 (Springer, Berlin, 2018)
2. L.M. Pecora, T.L. Carroll, Synchronization in chaotic systems. Phys. Rev. Lett. **64**(8), 821 (1990)
3. L.M. Pecora, T.L. Carroll, Synchronizing chaotic circuits. IEEE Trans. Circuits Syst. **38**, 453–456 (1991)

Index

© Springer Nature Switzerland AG 2020
E. Tlelo-Cuautle et al., *Analog/Digital Implementation of Fractional Order Chaotic Circuits and Applications*, https://doi.org/10.1007/978-3-030-31250-3

Printed in the United States
By Bookmasters